DEFENSOR FORTIS
By Kali Pinckney

Defensor Fortis:
A Brief History of USAF Security and Those Dedicated Few Who
Defend the Air Force at the Ground Level

Copyright   2003 Kali Pinckney

Universal Publishers/Upublish.com
USA   2003

ISBN: 1-58112-554-2

www.uPUBLISH.com/books/pinckney.htm

Acknowledgments

MSgt Kenneth Erichsen for all of the valued help and constructive criticism; Thanks for keeping me pointed on the correct azimuth. The assembled men and women of the Air Police, Security Police, and Security Forces that I've met this past 10 years. Edwards AFB, Ca and Aviano AB, Italy History offices, the personnel at the USAF Museum, the US Military Research Office, and of course, the Security Forces museum located on Lackland Air Force Base, TX. Without these people and organizations, there would be much less written Security Forces history than there currently is. Thank you for giving me some of the resources I needed to do my part.

The Blue Berets

The Army has their claim to fame
The "Airborne Ranger", the "Green Beret"
The Corps have their elite too
"Recon Patrol" the proud, the few
The world knows the Navy "Seal"
Life of danger and men of steel

But when it comes to us not much is heard
We're just the cops that guard the birds
In countless wars, through toil and strife
We give all our hearts, our blood, our lives
A page in history a chapter in time
We fight for freedom, yours and mine.

Always vigilant through the cold and rain
We bear the hardship, fatigue and pain
This piece of ground we shall defend
Side by side "till the bitter end"
So fear not during night or day
The base is guarded by the Blue Berets

Author Unknown

CONTENTS

The Military Police

During the American Revolution, the Continental Army was made up primarily of volunteer farmers. They were very undisciplined and untrained as far as their military duty was concerned. They were often drunk, fighting, and leaving the military areas they were assigned to without permission.

Then, as a General, George Washington's answer was to take an idea from the French Military and create a police force to keep his military in line. Duties for these first Military Police included patrolling and arresting drunken troops, thieves, protecting supply routes for other military forces, and carrying out executions. Military Police were also the valued and trusted guards of Future President George Washington himself. Defeat of the British ended the Presidents need for the Military Police.

During the Civil War, Military Police were brought back by the Union Army to keep its soldiers in line, enforce curfews, track the deserters, and most importantly, stop the plundering that was being committed by the Union Army troops against the civilian population. The Military Police primary mission was to support the front line combatants, although they did fight when duty required them to. Despite their value, after the war they would be eliminated once more, this time by Congress.

World War 1 began and the need for Military Police arose once more. This time, attempts were made to ensure that the force would remain after the war was over. MP Armbands would be worn and the 4 core competencies would be identified and established as actual written, and working doctrine. In this moment, MP's would gain an actual and real mission.

Their First mission was Maneuver and Mobility Support Operations (traffic and route control); this meant to help direct forces on their way to locations in foreign lands. Mission 2 was Rear Area Security, so the fighting combat troops would have less to worry about from their rear and only worry about the combat up front. Mission 3 was to maintain and control Enemy Prisoners of War (EPW) so that the combat units would be free to fight after turning EPW's over to MP custody. Their final Mission was to maintain law and order amongst the fighting troops, as well as investigate crimes.

Once again, after the war ceased, the Military Police were disbanded, this time by the Army itself only to be reestablished to serve once again at the beginning of World War 2. Recognizing that history had cycled MP's in and out of service for more 100 years, the Army would make MP's permanent, creating billets for, and training 2000 Military Police in 1941. By the end of the war, there would be more than 200,000 Military Police operating.

Enter The Air Force

The technology of warplanes came to fruition by the end of World War 2. The face and scope of combat was changed forever during this time. No longer could the aircraft be seen as an extension of a ground force.

Unlike World War 1, aircraft had gained the ability to destroy most targets on the ground through strafing and bombing. Aircraft were no longer mere balloons, dirigibles, or low powered bi-planes only useful in the visual identification of targets for the ground forces. Mere spying and engaging other low powered aircraft on the same types of missions became a relic of the past eras.

As the battle zones became three dimensional, new roles for aircraft affected and changed every other form of combat, whether on land or sea. Aircraft were successfully, and efficiently destroying tanks, trains, as well as decimating troop concentrations of the era. Aircraft were posing the most severe threats to shipping by sinking and destroying watercraft including the frigates and the mighty Destroyers that just years before were considered to be "unsinkable". After World War 2, the world would know the tactical and strategic power of aircraft.

Entire cities were destroyed by indiscriminate firebombing campaigns that set entire cities ablaze and caused hundreds of thousands of casualties. Nowhere was the threat of airpower more evident than the lethal combination of airplane and atom. An atom bomb of the "Little Boy" type was dropped on target and became the first nuclear device used in war. The bomb sent one sub-critical mass of Uranium into another sub-critical mass forcing a supercritical mass resulting in nuclear detonation (simplistically) destroying the Japanese city of Hiroshima. Nagasaki would fall victim days later to a slightly differing type of "Fat Man" type device.

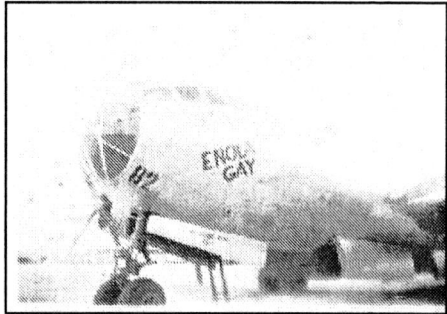

Done under the auspices of saving lives, both American and Japanese, it was decided to drop the bombs. On August 6, 1945, the B-29 named *Enola Gay* dropped an atomic bomb on Hiroshima, Japan. Almost immediately, an estimated 140,000 men, women, and children died as result. Three-quarters of the city (about 5 square miles) ceased to exist. Three days later, America dropped another bomb on

4

Nagasaki. On August 15, the Japanese surrendered. Unfortunately for history's sake, this weapon was perfect and worked flawlessly, just as developed.

The early 1940's was the period in which the idea for a separate Air Force was really planted. The last few years of WW2 made it all but obvious that a separate leadership might make for more effective ground as well as air components of the American fighting force. There was a working distinction between leaderships of the "Ground Army" and the Army Air Corps but that could be jumped fairly easily.

Army leadership at the time had to determine which was more important to spend funds on. The old school Army had been brought up on the belief that the armored tank was the king of the battlefield while other leaders realized that the aircraft had the full potential to be more effective than tanks in shaping the outcome of a conflict.

A more practical reason for the Air Force was to alleviate the saturation of responsibility on the ground commander so he could fight his battles. No ground troop commander

5

should have to deploy aircraft, plan his own air strikes, along with other tasks that do not help his direct surface fighting efforts, even if he thinks he should. At this point, combat aircraft were rightly considered support to the ground troops just as in WW1 where many aircraft would land and stage alongside the Armor, Artillery, or other ground units they directly supported.

The MP's Called Air Police

18 September 1947: The Air Force is born. The National Security Act of 1947 creates a separate United States Air Force that is not under the control of the United States Army, but an equal military service component. The foundation of the new Air Force would be the Army Air Corps.   All   personnel, material, equipment, and Military Police assigned to protect, maintain, operate, and otherwise support the aircraft and air bases became Air Force Property. Air Force MP's were basically Army MP's at this point in time, and simply assigned to the new air service.   There were no differences in training or employment as of yet. Brigadier General J.V. Dillon was the first (Air) Provost Marshal for the United States Air Force.   The Military Police shortly drew a

distinction as being in the Air Force as opposed to the Army. Even the name would be changed to "Air Police" to reflect this separation.

From 1947 through to 1959, the Air Police evolved from their Military Police roots of the few years before. During these years the Air Police were sorting out much of their own identity and traditions within the Air Force. The Korean War had begun to spool up and the career field grew by close to three fourths its size inside of one year. The field expanded from 10,000 troops up to 39,000 troops between 1950 through 1951.

Under the new Air Force Chief of Staff, the famed General, Curtis E. Lemay, the Air Police began to flourish. General Lemay was a big proponent of effective security as any great Chief of Staff should be. Under his leadership Air Police were now authorized to wear white "Bus Driver" hats to distinguish them from other Airman. General Lemay also presented the Official Air Police Shield to the Air Provost Marshal of the time, Brigadier General Burnham.

Air Police would soon become the

Security Police. The years of the Security Police era spanned from 1960 through to 1997. Most of the evolution of the current Security Forces obviously came during this 37-year period. Much of the rules that are currently recognized, from close in sentry posts all the way up to nuclear resource protection, have come from the Security Police regulations/era.

Security Police (SP) designation was established to reflect the change and direction to a force more capable at its two main duties; Security and Policing. At this point there was a split career field of Security Specialists and Law Enforcement Specialists.

The security for nuclear resources was of "Priority A" in the SP mission. Most other resources fell lower on the food chain. Aircraft Security and Air Base Ground Defense (ABGD) were all the secondary duties of the Security Specialist. Since most Law Enforcement Specialists were not generally required to enter Ground Defense training at the time, Security troops were the personnel used to protect the critical Air Force resources that were usually out of the site of the base populace.

Law Enforcement (LE) activities were also to be reflected in that Security Police name change. Policing of the air bases was likened to policing a small town. Most of the personnel on an Air Force Base are law-abiding citizens

8

that for the most part, were extremely
trustworthy and respectful. Security troops
were not as friendly and flexible when dealing
with average base populations. Law
Enforcement Specialist had more training in
this area of personal relations.

Base entry control was also usually the
domain of Law Enforcement Specialists. It was
a very important job that needed to be
accomplished effectively and professionally.
While the Air force has been a service, entry
control has always been of issue. This first
layer of security and crime prevention is
necessary to stop people who have no official
duty or authorized business on a military
installation from entering.

Very little change occurred over a 30-
year period within security because the mission
did not change. The cold war with the Soviet
Union consumed all of the military's resources
and defined purpose as only a sworn enemy of
the United States could.

1997 was the first year of the Security

Forces. This name change reflected the latest mission and the new direction of Air Force Security. No longer were ground defenders, base law enforcement (LE) patrols, and security personnel

separate little organizations. All efforts were intertwined into a singular, integrated security force capable of opposing whatever threatens the Air Force Mission. The emphasis is on counter-terrorism and Force Protection. Very few organizations find it beneficial to attack the US military head-on, however, small groups of trained aggressors represent a more likely threat to soft targets, including those on military installations.

Bird Watchers

There is one job that all Air Force security personnel are there to accomplish. Security Forces help insure that the aircraft can enter the fight by protecting them during their most vulnerable state. While aircraft are on the ground, they are in jeopardy because it is when they are airborne that they can run, hide, defend, or attack targets.

Security Forces personnel all maintain the basic knowledge of inflicting damage on aircraft as well as how to stop those intent on doing such damage.

Destroying aircraft on the ground is the easiest way to defeat an Air Force that cannot be stopped in the sky. According to the think tank, RAND Corporation in their study "Snakes in the Eagle's Nest", of 645 attacks on airbases, 384 (60%) were for the sole purpose of destroying aircraft and equipment while fueling, loading, and otherwise prepping for missions. There is a need for effective protection for aircraft. Generally speaking, aircraft are the target, not air base personnel.

Mobility places Air Force resources in a position that is usually not beneficial; such locations lack the support, services, and resources of large main operating bases. Most resources must be brought along to include resources for protection. Wherever the planes fly, security must go as well.

The missions are varied, from resource and personnel protection, contingency security, humanitarian operations, as well as direct combat. Aircraft generation is the key to everything the Air Force does. What the Security Forces do for the Air Force is provide an environment that allows all ground operations required to keep the aircraft flying safe and secure.

No matter what the situation, the key is the protection of aircraft; aircraft support equipment, and the personnel that maintain the aircraft thus allowing the aircraft to complete the tasks they are required to accomplish. Security Forces must always support the aircraft.

Operation SAFESIDE

1041st USAF Security Police Squadron

As the Air Force entered the Vietnamese campaign to the point where aircraft needed to be secured while in South Vietnam, Security Police went as well. Initially Security Forces were used to secure aircraft, never more than 25 feet away from the aircraft that were used to bring in supplies, support equipment, and military advisors to support the southern Vietnamese government.

Eventually, the concept of containment (of communism) would leave the support stage and enter the combat stage, which involved US troops directly battling the Northern Communists in the form of the North Vietnam Army (NVA) and the Viet Cong (VC). Aircraft needed access to land to bring in all the resources that would be needed to ensure US forces could fight and win a war.

The Security Police now needed to get away from the Aircraft and provide security outside of the base perimeter because that's how the enemy was attacking the Air Force. They were successfully using rockets and mortars to hit targets on base without risking their own safety. They would observe the base then exploit any weakness in security. Demolition teams (sappers) would attempt to penetrate base perimeters and destroy the targets that were not heavily protected. Many successful attacks included both of these methods used in conjunction.

A better and more highly trained force was now needed to meet the new threat, as well as test concepts that had potential to improve security on air bases. Problem was however, that there was no precedent. There was a concept of a standalone Security Police unit capable of providing all the support a unit would need to totally and completely protect an airbase.

The Air Force unselfishly provided the funds and allowed a unit to get the training that it required to do its job. The fact that aircraft were getting destroyed throughout the Vietnamese theatre definitely helped the Air Force in its decision to fund such a farsighted "experiment".

The project, code-named *SAFESIDE* started out with an initial cadre of officers and enlisted volunteers. This initial cadre first completed the US Army Ranger School at Fort Benning, Ga. After planning the training regiment and unit formation for the future 1041st was complete, this initial cadre met up with, and trained 200 more Security Police. A key factor to the success with the unit was that all personnel selected were volunteers.

The Volunteers were given 15 additional weeks of training in the Hawaiian Islands due to its climatic similarities with Republic of Vietnam. The training basically followed the model of the Ranger School that the cadre had previously completed. Completely physically fit, expert in weapons, highly effective in patrolling and reconnaissance. Many personnel were additionally trained in specialty areas including vehicles, investigations, intelligence, electronic sensors, explosives, demolitions, as well as forward air controller school.

After the training phase, the 1041st was deployed to Phu Cat Air Base, Republic of Vietnam. Their mission was to secure the Air

Base and test the Defense in Depth doctrine. Much of their success was based on Intelligence. There was not only on-base Intel gathering, but also off-base. Reconnaissance patrols surveyed particular points of interest as well as areas for any enemy activity. Electronic sensors were buried throughout the base and were very effective in giving the Safeside troops back the element of surprise.

Mobility was superior to most units. The 1041st had at its disposal Armored Jeeps that carried an additional 720 pounds of armor plating to aid in protection against not only direct gunfire, but also land mine detonations. M-113 personnel carriers were used to get the 6-man fire teams to the fight. The M-113 was used effectively to get troops around the 26-mile square base perimeter. Helicopters were utilized just as they were just about everywhere else in Vietnam. The 1041st utilized them to get troops to the more isolated, remote and normally inaccessible areas around base at a moments notice. The helo's also did well as support in reconnaissance, and aerial fire support.

Firepower was effective for this unit. Included in the Safeside "package" alongside their personal weapons were the heavy weapons. Four tubes of 81mm mortar provided the most powerful weapons support. With effective use, the mortar section provided illumination, harassment, and interdiction fire to support security efforts. Electronic sensors would pick up movement. Security troops would patrol out to investigate intrusions, verify the alarms, and immediately call for fire and attack targets aggressively with machine gun and/or mortar fire. Unlike most areas the United States military occupied in Vietnam, the number of alarms gradually decreased until it was practically nil.

Just as most other bases, Phu Cat had networks of tunnels and fortifications around base that were built by the Vietcong or North Vietnamese Army. When these were detected, unit members acting as Tunnel Rats would enter and investigate. After any possible Intel was gathered, the Safeside members who had been trained in demolitions would destroy the structure. In all, more that 350 tunnels, holes, or fortifications were destroyed.

Over a 6 month tour, more than 650 recon patrols and more than 150 Ambush patrols would be completed. The area secured by the 1041st accounted for both numerous kills and captures on Vietcong as well as NVA forces. There was not a single death and only minimal injuries registered on the side of the Safeside Security Police. Although the test would end, the legacy would live on through the Combat Security Police Squadrons and SPEC units that would follow.

Combat Security Police

CSP's were forces that grew out of the need to maintain highly skilled combatants in the Vietnamese theatre. These units were trained as Infantrymen from Basic Training on, and prepared to fight upon arrival to Vietnam. Many of the original members were former members of the original Safeside program after it was disassembled.

Combat Security Police personnel not

17

only included the SAFESIDE cops, but they also would grow to make up several units including the powerful and capable 82nd Combat Police Wing. Units fielded by the Wing included the 821st, 822nd, and 823rd CSP Squadrons. These unit's combat capabilities were based on the inputs of the Royal Air Force Regiments, US Army and Marine Corps on what makes for a competent fighting force.

During the Vietnam conflict, CSP's gained a favorable respect from other military organizations as well as enemy forces. Known for their professionalism as well as their aggressiveness and expert combat tactics, they would patrol the perimeters their bases in search of the enemy they knew was there.

Dependent on missions, some Combat Security Police wore the tiger-striped combat battle uniform of the time, usually supplemented by the use of one or two weapons which may include the Remington 870 shotgun and the .38 caliber pistol. CSP's were beneficial to the bases they were assigned and have a place in history as one of America's truly elite forces.

SPEC's

SPEC's were Security Police Element(s) for Contingencies. Established to both, supplement Air Base Ground Defense units that were stressed by combat in Vietnam, as well as to operate in "Commando" style security units that could take bare bases or critical facilities and secure them without much need for a large command and control or base support system that other Security Police units required.

SPEC's became necessary to fill the void left by SAFESIDE and Combat Security Police units as they began to disappear. The lack of financial support and lack of vision on the part of Security Police and Air Force leadership would win out once more.

SPEC units would perform normal Security Police duties on a base within the Continental United States (CONUS) until an emergency arose. They would then be deployed as an entire unit that would bring all the equipment and weapons they would need to accomplish an assigned mission. They were significantly trained and well practiced. Most units were driven by individuals that had previously gained combat experience in Vietnam. Upon arrival, they would be ready for a fight.

This idea of a contingency based, unit deployable, combat organization within the Security Police such as SAFESIDE and SPECs would not be seen until 40 years later in the form of the 820th Security Forces Group in 1997.

Security Training

Amongst the most effective times in the Career of Security Forces is when training. Although there is much time spent in classrooms learning the "paperwork" side of the job. Field times are just as valuable, but much more interesting in practice. Training includes practical combat training and field exercises that require Security Forces personnel to apply the procedures and tactics that they need to survive a given situation.

Training solidifies skills learned previously and gives troops a puzzle they must solve concerning Security Duties. The few moments spent chasing after an individual in training, tracking him down, slapping on the cuffs, and processing him into detention is much more effective training than sitting and trying to stay awake in a classroom learning the same skills.

Initial training in the Security Forces career field is adequate as a basic foundation but must continue to evolve in order to create well-rounded Security professionals.

Security Forces begin their track the same as all other Air Force Personnel at military Basic Training. For Air Force personnel, Lackland AFB in San Antonio, TX is the place they will go. After Basic Training graduation, security troops move across Lackland AFB to attend a basic security course where new troops learn some of the basics of security and military law enforcement. During this phase, weapons training is increased and additional weapons are taught. Courses in jurisdiction, apprehension, as well as military and police tactics are also emphasized.

After the completion of Basic Security Training; off to Camp Bullis, a United States Army Reserve Base less than 40 miles down the freeway from Lackland AFB. The next

phase of training security troops is to teach them how to protect and defend a base from other ground forces. Army bases are more suitable for the training needs required for Air Base Defense than most Air Force bases. From Camp Bullis, all the skill centers, firing ranges and training areas needed to train troops are already in place.

Ranges for tactical weapons firing such as running pop-up targets, up to heavy weapons can be supported; Every weapon up to and including grenade ranges. Miles and miles of woodlands, wetlands, land navigation compass courses, and MOUT (Military Operation in Urban Terrain) villages are ready for training use. Camp Bullis has such excellent areas that the USAF has spent millions of dollars to fund a Security Forces Compound for its troops to stay at that during their training that is second to none.

Training does not end upon graduation. Training will and should continue until the day of separation or retirement. Processes change

continually in Security Forces, so specialized training needs to be accomplished continually to ensure the success of a unit. Advanced courses on topics such as Anti-terrorism, Tactical Alarm Security Systems (TASS), Hostage Negotiations, or Vehicle Accident Investigations are examples of other courses some Security Forces will learn and fulfill.

Exercises are the most effective means of showcasing the capabilities of units and re-enforcing the training and practical skills of personnel in order to gauge how effective training programs are. Units use these exercises to train and give experience to troops that would otherwise never know how to handle a situation unless it actually took place, which is never a good time to learn.

Some of the larger exercises such as *"FOAL EAGLE"* in Korea and *"CRETE DEFENDER"* in Germany teach joint operations and involve forces from the Army conventional, and Special Operations forces communities, Naval Forces including SEALS, and Marines. Ground troops from other Special and conventional units from our ally countries also participate in such large joint exercises.

Peacekeeper and Defender Challenge

Peacekeeper Challenge was started as a competition between differing units and individuals trying to find who was the best Air Policeman. The original 1952 competition consisted of M-1 weapons firing, physical fitness, as well as other policing skills. Crime Scene processing, Military Working Dog, and accident investigations were included in the first Peacekeeper challenges.

"Peacekeeper" Challenge has since been renamed and evolved into "Defender" Challenge. Defender Challenge is a competition of combat related skills based on the Security Forces Air Base Defense mission. In addition to the regular battery of team and individual physical fitness events, Defender Challenge stresses tactical team movement and small unit combat operations. The firearms portions include numerous M-9 handgun, M-16A2 rifle, M-249 SAW, M-203 Grenade Launcher, and M-60 Machine Gun shooting events.

With all Air Force Major Commands represented, the competition has become International. Security Forces equivalents from Australia, Canada, and the United Kingdom are invited to attend and compete. Starting with more than 15,000 personnel worldwide, the final competition will end with a handful of winners.

The Tabbed Few

The US Army Rangers provide America units of well-disciplined, highly trained, and extremely motivated soldiers who possess the knowledge and courage to operate on their own, deep behind enemy lines. Rangers provide forward recon for regular Army units, immediate armed response, as well as utilize counter-guerilla fighting tactics to accomplish their objectives. For most special operating forces, ranger school is a prerequisite.

There has always been a connection between the Army's Rangers and USAF Security Forces due to the fact, the first Modern (Vietnam era) Security Police were Rangers. Besides any history; as a light infantry force, the model of how Security Forces fight is based on the Ranger's light infantry tactics.

This connection is not a formal relationship, but more of a young force with less tradition and direction looking for a role model and a goal to strive for. As the need for defenders to protect the Air Force arose, there needed to be that model of success.

To pick a role model, the Army as a whole was too vague, however picking and following the Rangers gave a more specific goal to achieve. They are a highly mobile, versatile, usually compact, small forces that can overcome larger forces due to training and tactics. They are very aggressive and make for an amazing role model to any military unit.

Strategic Air Command was a major proponent in the area of Security Police training. They were 100% in support of, and drove the areas of Air Base Ground Defense before the AF did as a whole. They even took the lead and were always attempting to get as many their troops as possible to and through Ranger school. As an example of this, SAC sent more that 1500 of its "Cops" to Ranger School over a one year period from 1970-1971.

Today, Pre-Ranger School is a must for any troop striving to go to Army Ranger School; The Air Force does not want to send an individual that would possibly fail to a course composed of Army, Navy and Marines making the USAF look bad when they quit. Many that have been through both suggest that the AF Pre-Ranger course is much more mentally challenging that the actual course.

You can train physical conditioning, you can teach skills, but you cannot create the mental toughness that Ranger school requires. The Air Combat Command (ACC) pre-ranger course at Indian Springs, Nevada boasts a near 100% pass rate at Ranger school. The people that go from there *will* pass as long as they are not injured.

This kind of pass rate is great, but may also serve to be failing the future of the Security. The Security Forces need to send more candidates to Ranger School. The more personnel trained to this level, the better impact Rangers will have in the future of the career field. The more Rangers, the more legitimacy, the better the future, and the better security will and Air Base Defense will be.

The Nukes

Some may argue, but it is said that the Air Force is the most powerful "single" organization within the world behind the Soviet Strategic Missile Forces. This argument is due to the fact that all of the Soviet Nuclear forces were placed under the single control of one command while the United States arsenal is broken into a triad, some of which belong to the Air Force and other's belonging to the Navy. This triad is based on a three-pronged strategy of nuclear stalemate. Of this triad, the air force controls two of the three components being the manned strategic bombers and the Inter-Continental

28

Ballistic Missiles. The Navy controls the other component in the form of its strategic submarine force.

The nuclear triad is the outcome of the most effective means of strategic defense for the time. The doctrine of Mutually Assured Destruction (MAD) made the act of nuclear war a sure means of catastrophe for the aggressor, and defending country alike.

The idea was that it was an impossibility to track and target all of the enemy's nuclear assets. If a country were to do a "First Strike" on its foe, it is a real and high probability that the enemy would respond in kind. This doctrine seems to have worked and can be considered a success on a strategic level.

Having the nuclear triad made the likelihood of diplomacy a much better option of conflict resolution. Each method of nuclear delivery had a possible weakness that could be exploited. Each also had definite benefits that the others lacked. Together, as a system, the triad was practically omnipotent. So effective was this doctrine that the Soviet Union mirrored and modeled it exactly, while adding and maintaining other components to insure they'd have an edge in a possible nuclear conflict. The Soviets added mobile launch vehicles based on huge all terrain trucks, as well as train based missiles that were all but impossible to track.

The Security Forces have always been responsible for the protection and final defense of the Air Force elements of the Strategic Bomber fleet and the Inter-Continental Ballistic Missile fleet as the first and last level of armed security to these highly destructive and priceless resources.

Submarines

The United States Navy's Nuke boats are perhaps the single most frightening wing of the American nuclear arsenal. They combine the idea of having an unknown craft within, or just off your borders, or anywhere in the earth's seas with fear of the precision and lethality of the Inter Continental Ballistic Missile. You know they are there and yet, there's not much you can do to combat the threat.

As the silent service, subs could be on site already hiding in any particular location at any time. They are (should be) Invisible to Radar, or Sonar until it's too late. Downside is it is possible to become targeted while in route by other submarine and sea based resources. Lots of nuclear payload that could be anywhere within a couple of weeks from the US is provided through submarine use. The protocol for the subs includes a launch of their payloads when they do not make contact with their commands after a prescribed time. So even if the United States was attacked and destroyed, an enemy should still expect an attack from the submarine fleets.

The United States Marine Corps is generally responsible for the physical security of the Department of the Navy Nuclear resources.

Bombers

"What good is a high technology, extremely high dollar, radar evading, stealth bomber if it can easily be destroyed while on the ground because of ineffective security?"

Bombers are the most versatile of the nuclear delivery profiles. The main benefit of bomber use is that they can be launched and recalled. They could be enroute by 12 hours or more, and be less than 20 minutes away from striking their targets all the while diplomacy is taking place. A Bomber can then be told to cancel their missions and return home at anytime before they release their payloads. This is a definite plus to any Government.

Downside to bombers is that they could be destroyed mid-flight even though they are really small targets in a large sky. Bombers may or may not pose a serious challenge to even an old fighter aircraft. Bombers can be

anywhere on the globe within hours with in-air refueling support. Decent sized payloads insured that bombers would pose a serious threat to any target. Strategic Air Command (SAC) was the command in charge of the AF nuclear arsenal for most of the Nuclear age until AF reorganization of commands caused SAC's demise in the early 1990's.

The idea of strategic bombing has changed very little from the Atomic Bombing of Hiroshima and Nagasaki. The main differences are larger explosive powers and greater accuracy of the weapons, and the superior capabilities of the aircraft. Low Observe-ability and Stealth allow faster, higher flying bombers to penetrate enemy airspace without being detected. This ensures long-range strategic bombers are a real threat in any engagement they enter.

Security for bombers is key. Bombers require lots of support and servicing. Time is needed to get them airborne for missions. Not only in refueling, but also on reloading times. It is a time consuming job; all the while they are on the ground vulnerable to attack by everything from a lone gunman to an airborne assault.

ICBM

Inter-Continental Ballistic Missiles are the last resort of the nuclear triad. Once they fly, something dies; once they are fired they are not coming back and they cannot be re-targeted in mid-air. The possibility that they can be destroyed after launch is so unlikely that it can be ruled out as an option until the Air Force can figure out how to do it. ICBM's take around 30 minutes to hit any point on the globe. To complicate the threat of nuclear weapons, Multiple Independently targetable Reentry Vehicles (MIRV's) replaced the single warhead concept and immediately made one weapon act as several by

giving the ability to destroy numerous targets along the missile's trajectory. Sea launched missiles also compounded the lethality and deadly effectiveness of the weapon's platform.

Thousands upon thousands of Intercontinental Ballistic Missiles (ICBM's) were on standby to lend support to any immediate response during the cold war between the US and the USSR. 1961 was the year of the first Titan Missile sites located at Davis-Monthan AFB, Arizona, Little Rock AFB, Arkansas, Malmstrom AFB, Montana, and McConnell AFB Kansas. Missile sites can be massive expanses that cover hundreds of miles from end to end and the Air Force is charged to ensure they are "All Secure" to this day.

"Northerntier" missile field duty, such as F.E. Warren AFB in Wyoming or Minot AFB in North Dakota eventually affects the vast majority of Security Forces personnel sooner or later. Although there is an ongoing push to limit and remove much of the nuclear arsenal, if a Security Forces troop is planning on staying with the Air Force for any length of time, Northerntier security duty is all but guaranteed for Security Forces. Most of this duty is pretty labor intensive and somewhat harsh based on the time of year. Zero visibility, 40 degrees below 0 temperatures, with a nasty wind chill in some conditions.

Security for the Air Force's nuclear arsenal is Intense. There are, as imagined, many and specific regulations to insure the greatest protection level possible. Not only is it hard to get within proximity to an American nuclear device, it would also be difficult to tamper with, and definitely to load, damage or release. For every single resource, there are units and posts manned by regulation to insure force dominance over most enemies.

Due to the vast size of the missile fields, security for nuclear resources takes its own life. Many skills were perfected and have been utilized over the years. Some notable areas are the use of camper teams to go out to the sites and provide security when sensors may be inoperable. Convoy skills are utilized on a regular basis to deploy and retrieve the ICBM's into and out of the missile field when they require maintenance that cannot be done on site. Helicopters and heli-borne (Air Assault) Security Forces are always beneficial in providing a gods

eye view of any action. Speed and mobility of the helicopters insure security personnel can be on any site extremely quickly if necessary.

In few other areas within Security or the Air Force are the rules and regulations so black and white, or the costs reach so high for failure. If a person crosses a red line here: this is what the security troops are supposed to do. If such an incident happens: this is whom the unit is supposed to call. And here is the appropriate response. If a person doesn't stop running after told to stop, this is what is what happens in order to stop them.

The safety of friendly forces and innocent bystanders will be considered, but will in no way deter security of nuclear weapons. If there were to ever be a question about security for a nuclear device everyone in proximity is subject to being shot, because of the political ramifications and risks to the world if one were to fall into the hands of unauthorized persons or organizations.

Security for nuclear resources had become far more effective. Newer technology, measures, practices, and storage facilities have limited and all but removed the opportunities for any unauthorized contact. Even though there have never been any major issues, better tactics have continued to develop anyway.

Security methods have proven that despite the tendency to believe in the flawed idea that "more is better", the ability and policy of "doing more with less" is possible when a plan is thought out, exceptionally planned, and expertly executed. Nuclear security takes advantage of force multipliers, training, and technology to provide better security with less manpower.

Through years of nuclear protection, there have been incidents, luckily there has never been an incident where true security or custody of nuclear resources has been lost. At the end of the day, Security Forces must assume complete responsibility and insure that nuclear weapons are secured and safe from any aggressor.

GLCM

Ground Launch(ed) Cruise Missiles (GLCM) were deployed by the Air Force in 1983. The GLCM program was an update and improvement on the idea of the German V-1 "Buzz Bombs" Rockets used during World War 2. Designed to send in a flying bomb that could hit its target within the enemy territory. This is a very efficient means of delivering ordinance; For GLCM, the payload to be carried was in the form of a thermonuclear warhead.

The Germans employed them as a random means of terror directed at the people of England, however, when used in another, "more responsible" manner, this is a relatively cheap means of devastating targets without risking the lives of pilots and expensive Airframes. The benefits included fairly long ranges, combined with extremely great mobility and reliability.

The GLCM program threatened as well as

terrified the Soviet Union by altering the way they targeted the United States nuclear resources. GLCM involved an ever-mobile Nuclear Missile threat that traveled around the European theatre.

GLCM deployed 16 nuclear missiles per mission and utilized large trucks to act as support, targeting, and launch operations for the missiles themselves. To the Soviet strategists, this meant that missiles were much harder to find and counter. In the US Strategy, this meant an ability to place some strategic weapons much closer to its enemy.

What the Russians saw was a severe, insurmountable threat that seemed more as a "First Strike" weapon. A weapon this close to it's target lessened the identification time that a country needs to determine if a missile has been fired at them and what they need to do in order to retaliate, thus negating the doctrine of Mutually Assured Destruction. Such an attack had the potential to kill the authorities that authorize the launch of defensive missile attacks.

During the program there were 464 GLCM's in use across Europe, and deployed out of six Main Operating Bases (MOB). The six MOB's were (1) Greenham Common in the United Kingdom, (2) Comiso in Italy, (3) Florennes in Belgium, (4) Wueschheim in

Germany, (5) Wonensdrecht in the Netherlands, and (6) Molesworth, also in the United Kingdom.

The Soviet Union protested the GLCM presence, and for good reason. The US had parked and traveled Nuclear weapons less than an hour away from its country. The Missiles could travel up to 1,500 miles at more than 550 miles per hour below radar detection altitudes. From GLCM tracks in England, much of the Soviet Union, including Moscow, was within range.

The United States was able to get concessions in the form of a new Intermediate (range) Nuclear Force, or INF, Treaty which was signed in 1988, and ended the GLCM program. Although it ended, there can be no doubt that it was a true strategic success in bringing the Soviets back to the nuclear bargaining table.

The missile itself was the derivative of the US Navy's Tomahawk cruise missile program. Tomahawks were simply adapted to fire from a new land-based platform. Practically all of the changes were internal, which involved updating and re-writing much of the software that would make this missile work from its new launch vehicle.

The launch vehicle was designed with an erector that would hydraulically point the missile skyward. The vehicles themselves

41

traveled with the full compliment of support services to travel its route. Provided was everything it needed to ensure security, food, communications, and anything else that could be imagined for any and all situations including the case of a launch of its payload if that were necessary.

Security Police that traveled with the GLCM had to be prepared for anything. This assignment included much running, training, and practice to master. Advanced combat skill training was required. As the convoy moved, the normal level of high security for Nuclear weapons was adhered to, even in such an uncontrolled environment.

The constant convoy movement, digging-in and packing-up cycle was ever present. Foxholes were dug, radio's set-up, sites were camouflaged every time the resources stopped. After minutes or hours in one location, based on mission profile, the convoy would pack up and move out. The cycle would start over.

GLCM could at times be extremely grueling Security Police duty, but it was real! You would be hard-pressed to find a former GLCM Security Troop that did not think he was doing something important and vital. There was a real threat, and a real actuality that those missiles would fly someday. Personnel involved with the program will constantly echo the reply that GLCM was their best ever Security Police duty.

Air Base Defense

As soon as the Air Force became its own service, the issue of who would protect Air Bases arose. The Army Infantry ground forces inherently stationed on Army Air Corps bases prior to 1947, and prior to the Air Force, were now gone. All Army personnel went through a basic training program that introduced them to ground combat skills and taught them how to fight while the Air Force swiftly ridded itself of basic combat skills training.

With the exception of menial service jobs and simple duty, the US military had a policy of segregation. The United States had a history of calling on colored (non-white) people to fight when there was a conflict or war, only to immediately fall back to segregation when necessity was gone.

President Roosevelt took initial steps to satisfy some of the complaints of colored (non-white) Americans in June 1941 when he issued Executive Order 8802. What E.O. 8802 did was prohibit discrimination in "employment of workers in defense industries and in government because of race, creed, color, or national origin". The tide began to turn for blacks and other minorities within the military.

The fight for desegregation would continue into the mid 1950's spurred by President Truman's Executive Order 9981, which directed "equality of treatment and opportunity for all persons in the armed services without regard to race, color, religion or national origin".

Air Base Security Battalions

The military hierarchy still held on to its segregation until it could not be avoided anymore due to threats from the then, Secretary of Defense. Although George C. Marshall had to be "pushed" to integrate blacks within the ranks, he made history by ensuring that the Air Force was the first service to complete integration.

Just as the Tuskegee Airman, during the same year, the first all black "Defenders" contributed to the history of the Air Force. Opportunities to do real jobs within the US Military insured immediate enlistments. The vast majority of the 53,299 blacks allowed to enter were directed into the role of Air Base Defense. In the Air Force, no doubt this was a means to "follow orders, without actually following orders" by the Air Force leadership. Blacks were officially integrated into the service, but not really integrated into many of the jobs considered "important". It would take years to truly equalize.

How this affected security was by

providing a fairly large group of bodies that had to be given jobs. The Air Base Security Battalions were all black enlisted units with all white officer corps and structured on the RAF Infantry Regiments. Despite problems of funding, racism, and lack of training, the units did particularly well with the resources they were given.

More than 20 (22 or 23) battalions were trained and dispersed to the four winds. Almost all units saw action in North Africa, Italy, and throughout the Pacific Theatre. The history of these troops is a very important story in the Security of the United States Air Force as well as the acknowledgement that the Air Force was recognizing the threat from ground attack.

The Air Base Security Battalions (ABSB) at this time operated the most firepower of any era of Defender, to include today's force. Amazingly enough, the Security Police of the Korean or Vietnam wars could not even compare with the combat power an Air Base Security Battalion could direct. Along with their personal pistols, rifles, and grenades, each battalion was armed with an effective supply of .30 and .50 caliber heavy machine guns. Most impressively, they were armed with four M-2 Half-track trucks, twelve M-3 armored cars and four Self-propelled, 75-mm, artillery guns.

Fencelines and Perimeters

Air Force security leaders continually debate "do air base security personnel stay behind the fenceline and stand on the planes waiting for attacks or do they get proactive by patrolling and extending beyond the confines of the base perimeter"? Obviously, a base would be more likely to survive an assault if detection range is increased and preparation time is used properly. In defense, the element of surprise must be denied, and this cannot be done without patrolling off-site in one form or another.

The United States Military does not historically recognize that Air Bases are Key Terrain as opposed to rear support areas. Key Terrain is defined as, that real property that must be denied to the enemy because it's loss of control could be utilized to the detriment of the mission. World War 2 saw numerous assaults on such key terrain, whether by conventional forces, airborne troops, special operating forces, and even by blitzkrieg. Air Bases will continue to be prime targets in the future, especially by forces hostile against the United States that cannot hope to win against US Air Power in the sky.

In 1950, General Curtis Lemay ordered Provost Marshall Luper to build a ground combat force for the Strategic Air Command (SAC) after one of SAC's bases was attacked. The force was to be highly trained, and expert by any military standard at the time. This force was the precursor of modern Air Base

Defense and the troops that participated were nicknamed "Lupers Troopers".

The British Air Regiments are a sort of cousin to the Air Force Security Forces and they still field dedicated units for the protection of their bases just as the role Air Base Security Battalions did. Raised in the same era as American air base defense, the Regiment is a well planned out force that has the capability to back up the other British services. Imagine the Air Force supporting and participating in operations with the US Army or US Marine Corps on a ground engagement if reinforcement personnel were needed quickly. Not many Security Forces, or sister service Commanders would feel comfortable with such an arrangement for anything this side of an emergency.

The British Regiments are composed of extremely competent combatants and are trained on par with their Army, and Marine brethren. Exchange Officers have been a benchmark in the USAF relationship with the British Regiment for more than 20 years. There are Regimental Officers at the Security Forces Training Academy that add their expertise to the U.S. Air Force. The Regiments have the capability to defend the Royal Air Force (RAF) from a varying and well-trained enemy.

Throughout the History of the Air Force, there has not been one clear-cut doctrine as to how to defend an air base, which gives a Ground Force Commander great versatility in his defense. The tools are there and the basic concepts are taught at various Security Forces Levels, however much is left to the discretion of the Security Commander how he will protect his base. Some Commanders will protect bases better than others. One rule is constant however. "Those who try to detect the enemy farther away from the base are providing far more effective security and defense for their areas of responsibility.

Key West Agreement

In March of 1948, Secretary of Defense, James V. Forestall called a conference to discuss and set service functions with his military leaders. His goal was to clarify some of the duties and responsibilities that each branch of service was to play in combat operations in relation to the other services.

There were several areas of contention; in particular was the question of whether or not the Navy had a role in strategic air operations. In other words, should the navy be able to get funding for large cargo aircraft and strategic bomber aircraft. The answer would be "No", as this was an Air Force duty. Other areas of the KWA limited the size and role of the Marine Corps as to not enter the Army scope of operations.

A major area of concern to the Air Force's Security was the question of "who should assume primary responsibility for land based air defense and, how would the Army aid in defense of air bases" since the Air Force did not want an entire "Infantry Force" for it's own defense as did the British Royal Air Force.

The Key West Agreement (KWA) is so pivotal for Air Force Air Base Defense because this is most likely the document that normally intelligent personnel refer to when they say that it is the Army's responsibility to defend Air Force Bases outside of the Air Base perimeter.

Although general responsibility for ground operations throughout a combat theatre falls to the Army, there was no magic number to say that 3, 5, or 20 miles outside of the airbase fenceline is the Army Line Of Responsibility. This was not established.

Whatever the case, this is up to the Defense Force Commander to establish when he arrives in a theatre and can only be established on a case-by-case basis. History does show, however, that the Air Force has always played the major role in its own defense. It must always be prepared to continue that tradition for future conflicts. The Air Force must be prepared to protect itself, just as it has needed to do in "every" conflict since it has been an Air Force.

Korea

Of all the locations that Security Forces personnel are stationed. There are a few hubs that large numbers of Security Forces seem to "Visit". Those are the Missile fields in the northern United States and the bases within the Republic of Korea. Korea's bases are a key to peace and stability in that area of the world. The main air bases include Osan and Kunsan; both bases are key to defense in the event North Korea was to attack South Korea. There are also several geographically separated units throughout the region.

After World War II, a republic was set up in the southern half of the Korean Peninsula while a communist-styled government was installed in the north assisted by Russia. The Korean War lasted from 1950 through 1953 and had US, and other UN forces intervene to defend South Korea from North Korean attacks. At one point, the Allies were all but guaranteed victory until the Chinese entered the war in a direct capacity and overwhelmed American and allied forces with a million Communist Chinese soldiers.

An armistice was signed in 1953 splitting the peninsula along a demilitarized zone at about the 38th parallel, which is the current border. With troops stationed on both sides, there have been skirmishes and incidents over the years, however no incidents grave enough to reinitiate combat. As a beacon of democracy, South Korea achieved rapid economic growth, with per capita income rising to 13 times the level of North Korea.

The significance of Korea to the Security Forces is that it was the Air Force's first major war and the site of our initial failures in Air Base Ground Defense. Most have forgotten the important lessons of this terrible war. Security Police lost resources during this war. Not simply to the point of losing aircraft, but to the point of losing installations.

Bases were completely and totally overrun. Kimpo Air Base was overrun on June 1950 and combatants were killed in action. Personnel who didn't fight back and surrendered were taken to a large aircraft hanger and executed by North Korean forces.

The reasoning behind these devastating losses was that there was no training on how to organize, and defend bases. Air Base Ground Defense was not a big priority for Security Police units of this time.

The Korean frontlines were brutal and the Army had a mission to accomplish and needed to utilize its forces effectively. This left the Air Force with token base security patrols and the occasional aircraft security team. This was merely a show of force for the attackers. The weapons may have been there, but practice and training were not.

Because of the possibility of combat in Korea currently, and the state of war that continues to exist, training is emphasized and supported by all units. The weapons and resources are for an actual threat. Armored personnel carriers are still utilized as well. The Stinger Missile system was utilized to help ensure that there is a capability to destroy aircraft present by ground defense forces.

The lesson that should be taken away from the Korean War is that the Army has its own objectives it needs to achieve as well. When it meets resistance around their fronts, amongst the first troops pulled to resupply and act as reinforcement forces are the troops that are charged to protect rear areas, including Air Bases. The Air Force must understand and have a capability to protect itself.

Vietnam

"I expect that our combat battalions will be used primarily to go after the VC and that we will not be forced to expand our capabilities simply to protect ourselves". "All forces of whatever service who find themselves operating without infantry protection... will be organized, trained, and exercised to perform the defense and security functions. "

-Gen. William C. Westmoreland, 1965

The Republic of Vietnam has a great history. Not only was it once considered the Paris of the Orient, with beautiful architecture, historic castles, fortifications, and amazing food. Culturally, it has had much to boast about. Once the prime location for French vacationers, it has retained many French traditions and building accents. Tourists from around the world regularly traveled to Vietnam for the luxury and beautifully exotic countryside.

Vietnam also has a long history of

fighting for its cultural freedom and identity. First the Japanese went in and the Vietnamese fought them out. The French came next and they were also fought off. Then came the unfortunate battle between Democracy and Communism. American leadership felt that if Vietnam fell to communism, other countries in the region would fall to communism as well. Just as dominos fall so would Cambodia, Laos, and Thailand, and this was unacceptable.

America would enter a war to halt the spread of communism on Vietnamese property. Obviously, America would win militarily, but it did not win the real war of keeping communism from spreading. America would end up leaving just as the Japanese and French did. Many thousands of American Forces would die, but hundreds of thousands of Vietnamese would die during the same period.

Much of the knowledge and expertise that Security Forces know and accept as the gospel was gained from this conflict. Security Police absolutely excelled in defending bases. There were personnel losses; however, the ratio of Security Police to enemy combatant losses was outstanding.

Experience from the Vietnam conflict shows us that no matter who is primarily "supposed" to protect an air base, Security Forces of the Air Force should (in doctrine and mission) assume ultimate responsibility for protection of the aircraft, personnel, and equipment on air bases. Unfortunately, this is one of the same lesson that was learned and forgotten from the Korean War.

Research done on Vietnam by Alan Vick, in his benchmark work, *Snakes in the Eagles Nest* (RAND Corp) shows various points of interest. However, three facts are undeniable. (1) During Vietnam, air bases were attacked 475 times, (2) 99 aircraft were destroyed in those attacks, and, (3) 96% of attacks on Air Bases were done with standoff weapons in an attempt to destroy aircraft off site, as opposed to sending troops through base defenders in order to destroy the resources.

Base attacks in Vietnam were used primarily to shut down operations and destroy aircraft. Enemy forces never attempted to take over airbases for their own use. Vietcong and NVA forces did not have the technical expertise to operate USAF aircraft so their capture would be of little use and cost too much in troop lives. Destruction however, was priceless for insuring victory for the north since aircraft were demoralizing, causing losses, and simply posing a threat to the Communists forces free movement.

99 Air Force aircraft were completely destroyed while on bases, however, over 1000 more aircraft were damaged while on the ground. Of that, more than 500 of them required repair and inspection prior to their next sortie. This was definitely at a cost to Air Force resources and operations.

Far more aircraft were damaged or destroyed on the ground on American air bases than were ever damaged or destroyed in the skies in over Vietnam. Based on the statistics alone, you would surmise that if more money had been spent on defense and security, better defense and security would have saved a few of these aircraft. The cost of one or two of these destroyed aircraft could have easily provided for far better base security all around.

During the war, the enemy quickly realized that they were not going to stop the Air Force by going head to head with the defenders. Defenders have an inherent advantage over the attacker. Not only do they have time to prepare and practice defenses, they also "should" know their area of operation better. Defenders "should" also recognize what is normal and what is abnormal within their areas.

Security Forces units understand that the charge of base defense is their specialty, that theirs is likely the last line of defense, which is perhaps the greatest motivating factor to air base defense. This should not simply apply to the individual troop on a post, but also to the unit as a whole. Security needs to be proactive and aggressive in detection and destruction of threats outside of base perimeters. This is why combat patrolling, listening posts and observation posts (LPOP), and Short Range Reconnaissance Patrols (SRRP) are vital to the future of airbase security and defense. Early detection of the enemy and the element of surprise determines who wins or loses when Air Base Defense is concerned.

The Air Force had no sooner forgotten the lessons of the Korean War and it took the first few battles and loss of aircraft in Vietnam to help jog memories on what was possible and what could happen if security was very low priority. An example of this is that in 1964, Tan Son Nhut air base defenders had whopping six vehicles for a 16 mile perimeter. In no-ones book is this acceptable. Towards the end of the war and the losses of numerous aircraft, priority was increased and funding sprouted for items such as remotely monitored radio alarm systems, armored vehicles, training, detection dogs, and weapons platforms. Most importantly, the SAFESIDE project was funded. Security was improved and fewer aircraft resources were destroyed towards the end of the war.

1969 brought the first deliveries of M-113 and M-706 vehicles. There were more than thirty M-113's in use by Security Forces showing that the Air Force as a whole was getting the message that they needed to protect themselves. After realizing that some of the rocket attacks were hitting close to home, Base Commanders began to realize that their personnel, resources, and equipment, including aircraft could become a casualty of attackers so they pushed to get more funds and resources into the hands of base defenders.

Vietnam had 10 Main Operating Bases (MOB's) that Security Police were actively trying to secure. In addition there were several Forward Operating Bases (FOB's) including some classified radar and supply sites in countries that would not have approved. All in all, 99 aircraft were destroyed while on the ground and more than 1000 were damaged.

Security Police also have the distinction for the last "official" combat engagement of the Vietnam War. As America was running 24-hour refugee flights, a sniper started taking shots at transport aircraft. The sniper was shot and killed by Security Police assigned to aircraft crew duty (reminiscent of Raven) just before they loaded up and left Vietnam.

The Tet Offensive

Tet is the most important holiday for the Vietnamese. It is actually a Lunar New Years celebration that calls for all people to start the new year right. Houses are cleaned and even repainted, clothing is bought, and a trip to the Pagoda (church) to pray for happiness and good fortune for the new year. Tet lasts for 7 days and little non-essential work should be done.

One of the most aggressive enemy attacks took place during Tet of 1968 by the 3rd Security Police Squadron. In a period where conflict is usually put on hold, the enemy, composed of North Vietnam Army and Vietcong, exploited the expected truce and attacked US Forces across the country of Vietnam. Air Bases were not exempt.

One of the most extravagant battles that Security Police were engaged in took place at Bien Hoa Air Base. The attack was initiated with a long rocket and Mortar barrage as they usually did. At the same time, enemy forces penetrated the base perimeter. Military Working Dogs alerted and the gun battle erupted. Security Police did their best to defend the aircraft and the perimeter.

There were no crew-served weapons only small arms possessed by the unit, the largest being the M-60 machine gun. Additional equipment was also sparse including no Armored Vehicles, only Jeeps and Pickup Trucks. All bunkers stood strong and none were overrun. The only support afforded to the Security Police were a few Army of South Vietnam (ARVN) troops with a few recoilless rifles, Air Force augmentee personnel who were to stay on the inner perimeter, random Army and Marines who were transiting through, and most importantly, the US Army's 145th Aviation Battalion and their Helicopters.

The 145th helicopters were invaluable for their Close Air Support. They spotted troop movements, initiated suppressive fire with their door gunners and most importantly, re-supplied Security Police on the ground with ammunition so they could continue to fight. The Army was effectively providing Close Air Support (CAS) to the Air Force.

After battling all night, the dawn brought an Army Tank and 8 Armored Personnel Carried that would chase the repelled attackers back into the jungle. The base was again secure. Prior to the attack, there were no Army units patrolling the perimeter despite the Air Force being scolded to keep its personnel inside the perimeter.

In the end, Security Police would lose 4 (KIA) and 26 (WIA)[1]. The Enemy lost 126[2] of its

---

[1] One record indicates 2 Killed In Action (KIA) and 21 Wounded In Action (WIA). Author was unable to confirm any further.
[2] Or 139 Enemy KIA depending on sourcing. Author was unable to confirm any further.

personnel and 25 captured by American Forces. There were also 24 other friendly forces killed to ensure this battle was won. Battles such as this were played out similarly at the other Air Bases Security Police manned all across the Republic of Vietnam.

Duty in Vietnam

As it was for all troops, Security Police faced rough environmental factors, including extreme heat, insects, and of course, the ever present threat of combat-related injuries. As always, a large number of SP's thought they were being used wrong and continually pushed the envelope on their duties, for their survival.

Many Security Police wanted to get into more action and would go on combat patrols around their base perimeters even though they were warned against this continually. Some would even go on patrols with the Army or Marines units when it could be worked out.

Some Security Police served as helicopter door gunners on their days off to "help the Army out", so their crews could get days off. Such activities were obviously not authorized, and repeated threats to cease these activities were given, but many blind eyes were turned.

It would be fitting that Vietnam would be such an important war for Security Police. Of its almost 60 years in existence, 17 years were during this war. Modern Security Forces have a direct relationship to the Security Police of that era.

USS Mayaguez

On 12 May, 1975 the an American ship named the SS Mayaguez was on a regular shipping lane when several small PT (swift) boats surrounded it. One of the small boats fired a 76mm round across its bow. The ship was forced to follow the PT boats to Koah Tang Island, which is roughly 30 miles off the Cambodian coast.

On May 13th The first casualties of the effort to free the SS Mayaguez from its Khmer Rouge captors would be a CH-53 helicopter full of Combat Security Police from the 56th Security Police Squadron. According to the plan, 50 troops would assault the ship by air and secure it by force. Helicopters would transport the troops and release them over the shipping containers. The Security troops would recapture the ship.

While enroute however, one of the helicopters disappeared from radar. It either suffered mechanical problems or was shot down and crashed into the dense jungle. 18

Security Police and the 5 member Special Operations helicopter crew would die. After the crash the, remaining Security Police force returned to their base and a Marine detachment was floated in to secure the ship.

Had these troops made it to their destination and successfully completed their mission of securing the ship, history would have went down as saying that Security Forces of the Air Force are capable of being used during offensive combat operations. Security Forces leadership might be more inclined to provide effective training in combat oriented areas today. Sometimes there are pivotal moments that determine the future.

These Security Police lost their lives in an accident, but they should be remembered as the warriors they were. Years after the controversy of how and why the craft went down, all members were Posthumously awarded Bronze Stars (with Valor) for making the ultimate sacrifice.

Female Security Troops

Throughout the history of military service, it has been traditionally thought that fighting and being in the military is "man's domain". Subsequently, as females have found their way into the military, there have always been jobs that have been left to men and men alone. Today, the last holdouts remain in all of the service's Special Operations and Special Operations Capable Forces as well as some extremely exclusive jobs such as Navy Submarine duty.

It would be disrespectful not to the mention the introduction and role of females to Security Forces. Sexual segregation was included within Security Police and Air Force history, just as it was within the Army, Navy, and Marine Corps. Females were banned from holding a position within the Air Police and Security Police until the rules were changed to allow for their service.

Prior to 1971 there were no females in the two career fields that made up the Security Police. There was the Security Specialist and the Law Enforcement (LE) Specialist. They were trained separately and worked differing duties. In November 1971, the first 6 females graduated from LE training. LE was considered to be less physically demanding, and civilian police departments were proving that females could do the job with little problems. They were correct.

August 1973 brought the first female Dog Handlers to the Security Forces. Groups of these initial females were trained and sent

together to their first duty bases. It was tough duty with the established sexism inherent in the vastly predominant, male career field. Despite all the trials and tribulation, much of which was directly ignored by Chains of Commands, women would succeed in Military Working Dog units.

As more and more females entered the ranks of the US military, there became a need to house increasing numbers of them into the USAF/DOD Corrections and Confinement systems. 1974 was the first year of the female Corrections course graduates. There were not many females in the jails, but the services worked together to house as many of their females together no matter what branch they originated from.

In Mar, 1976 2Lt Pam Kraus was the first

female officer to graduate Air Base Ground Defense training at camp Bullis, New Jersey. She was the catalyst that proved that there was no real problem for females successfully completing this type of training. Following 2Lt Kraus' success in Nov 1976, 100 enlisted female volunteers were selected for Security Specialist training as a test for females in combat related jobs.

Due to many reasons, the main being a sexist attitude that prevailed within the entire Security Specialist career field at the time, the test was considered a failure even though the vast majority of females excelled. The times could not be held back and in 1985, females returned to the Security Specialist career field and this time they were/are here to stay.

In January 1985, one female (Airman Queen) of the Air National Guard would graduate Security Specialist training. Twelve females would become Class 850510, the first all female flight six months later.

Airman Laurie A. Lucas becomes the first Security Policewoman to die in the line of duty in August 1991. During a standard force-on-force amphibious (landing) training exercise with the US Marine Corps. Amn. Lucas manned a M-60 Machinegun. She was in a elevated position and had the situation been real, would've caused considerable casualties to the advancing troops. A Marine tossed a

Ground Burst Simulator (simulated explosive) up into her position, where it exploded, mortally wounding Lucas.

Today, the military has insured that over 98% of its jobs are open to female candidates. Women have done an excellent job and proved that much of the worry was unfounded. If there are quality candidates and all standards are met, all should be allowed to progress. Security Forces females are no different.

Tools of the trade

There have been hundreds of weapons configurations and platforms that the Security Forces have been trained to operate over the years. These weapons have been utilized in specific instances throughout Security history. From various submachine guns, to sniper rifles, shotguns, up to bazookas.

From .22 caliber rifles, tranquilizer guns, all the way up to 90 mm recoilless rifles, there have been perceived uses for all types of weapons. Some examples include the "Grease Gun" and the always fun, Israeli made Uzi family of weapons.

The Carbines

The standard issue rifle that the first Air Police and Security Police carried was the .30 Carbines. The M-1 and M-2 carbine rifles were in wide use throughout the military during and after WW2, the Korean War, as well the early stages of the Vietnam War. The gas system was actually designed by David Williams. Ironically, Mr. Williams was in prison for murder at the time he designed and created the revolutionary gas system. Winchester Arms bought the weapon and hired Williams after he was released back into society.

Williams worked with the Winchester machinists to design a prototype to submit to the U.S. Ordnance Department for testing. The Ordnance Department adopted Winchester's

design of the M-1 Carbine in October 1941 and every branch of the Armed Forces utilized this weapon during World War II. The weapon was still in widespread use during the Korean War and even early on in Vietnam.

The M-1 was successful as a rifle, however it was time to upgrade to a better rifle. The M-14 rifle was a 7.62x51 (308 cal). rifle that would be much more powerful than the M-1 carbine and had a flatter firing trajectory and a larger projectile. But, although some M-14's would see service with the Security Police, it would actually be the M-16A1 that would win out as the new standard issue combat rifle.

Operation: Semi or Full Auto, Selected by switch added to left side of receiver, operating on the sear mechanism
Length: 35.65 in.
Weight unloaded: 5 lb 7 oz
Magazine: 15 or 30 round detachable box
Ammunition: 108 grain bullet 30 Cal.
Effective Range: 340 meters

Colt 1911

There have been numerous weapons made over the past 200 years; however, few weapons will go down as legendary. The Colt 1911 series of .45 caliber handguns will go down as one of the best automatic operating handgun designs ever. The pistol had the right action, the right design, the right idea, and most importantly, the right bullet.

The .45 caliber bullet is still considered a benchmark of performance to this day. Many a military veteran and civilian police officer swear by this round and if given a choice between this and a 9mm there is no question many would chose the "45".

The '1911 is a great sidearm, but its technology began to show its age. Although it is proven and has stood the test of time, it was best to be replaced when it was. One of the main arguments that people have about the benefits of the '1911 is the supposed "stopping power" the weapon posses. Stopping power is the heavily debated ability to stop a target effectively. This is the, one shot will knock the target down, argument that is an issue unto itself, but the fact remains that you need to hit what you shoot first and foremost.

The .45 was most effective with shooters with larger hands and the opportunity to fire off plenty of training rounds. The 9mm automatic that would become the standard military pistol in the mid 1980's by contrast would be easier and more accurate for the average shooter.

Although there are many modernized and upgraded '1911 "raceguns" out there that reinforce the fact that these handguns can be made to be extremely accurate, the standard military versions of the weapon had much to be desired and were not really known for their accuracy. The current 9mm pistol fills a wider range of hand sizes more sufficiently and allows a more varied and average quality of shooter to be effective.

.38 Revolver

By 1960, Basic Air Police training would include the employment of the .38 caliber pistol with a revolving magazine. Not until February 1962, however, does the revolver make its introduction to the base level security.

The Air Force, as a service, had never used revolvers before this as the Army, Navy, and Marine Corps had in their histories. It was forced to adapt and implement the sidearm as standard. Much of the decision to implement the .38 cal is due to studies and experience of civilian Police departments as a viable and effective weapon for law enforcement use.

As far as many Security Police see, however, is that this is a step backwards after using the Colt auto pistols. Other reasons for switching to the 38's include the aging weapons owned by the Department of Defense and the cost of operations with the .45 Calibers. The costs to maintain the "fleet" of weapons became astronomical due to breakage, age, and wear.

The costs and benefits to implement the new revolvers on the other hand were very attractive to the people in charge of the money. Not only was a .38 cheaper to buy than the .45 semi-automatic, but simplicity insured far less parts to wear out, which required far less maintenance time, which made it less expensive to train troops on as well as keep trained due to the cheaper cost of sending ammo downrange.

Confidence of the troops in the weapon was minimal. Security Police command refused to listen to the experiences and concerns of the troops actually fielding the weapons. The most important concern was that of the time required to reload the weapon. In combat, or a firefight, it would be a very likely possibility that a troop could get caught with his pants down in the middle of a reload.

The people back in the offices and in charge of the money disagreed.

Currently, the entire remainder of 38's are relegated to training, most being utilized for training Military Working Dogs. This is the perfect use for a revolver since it can be used to shoot blank rounds around personnel or dogs without requiring modification to the weapon.

Remington 870

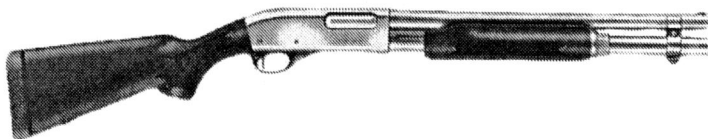

Shotguns have always had a place in the Security Forces just as they have in all law enforcement agencies. Nothing quells an argument or disperses troublemakers more than the sight of shotgun; or sound of a shogun being charged. Even people who know nothing about firearms recognize the distinctively abrupt double click of a pump-action shotgun.

The main use of shotguns is in the case of tactical situations such as barricaded suspects. Other uses include stray animals and in the event of an aircraft crash in a remote area, protection from wildlife.

During the Vietnam War, the Combat Security Police made excellent use of the shotgun on patrol. Many patrols through the dense jungle around the base surprised and terminated enemy troops with Remington 870's, which coincidentally are still the shotgun of choice for security units worldwide.

Current use of the Shotgun is mainly

based on the various tactical teams throughout the bases. Many configurations have been added such as folding stocks, side-saddles for additional ammunition, or flashlight mounts. The original sights may be replaced by rifle-style, ghost-ring (peep), or even lasers sighting systems.

Security Forces do not regularly need to utilize shotguns during normal day-to-day operation, but just about all units maintain several just in case the need arises.

Type: pump-action, tilting breechblock
Rounds: buckshot, slugs, tear gas grenades (Ferret), less than lethal rubber bullets
Gauge:12
Barrel length: 12 or 14 inch
Capacity: 8 round with tube magazine
Weight: Varies with model year, stock type, and barrel
Length: Varies with model year, stock type, and barrel
Maximum effective range: 50 meters

M-79 Grenade Launcher

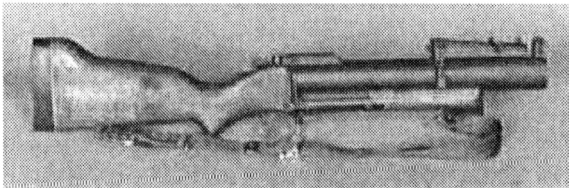

The M-79 was the first widely successful grenade launcher in use by the United States Air Force, and US military as a whole. The '79 is a stand-alone, manually operated, breech loading, launcher tube that was surprisingly accurate in its simplicity. The 40mm grenade traveled "accurately" out to its maximum range of 400 meters. This is of course, based on how well the grenadier can judge (guestimate) ranges.

A few M-79's still find use in today's Air Force in specialty roles. In training roles, the M-79's are used as a means to deploy smoke grenades and flares. Some of the more efficient Security Forces units will still have these available for riot control formations. Just as in the civilian sector, M-79's have been primarily relegated to the tactical teams. This weapon has been a favorite of civilian SWAT teams that utilize them to deploy smoke, tear gas (CS/CN), and even beanbag less-than-lethal rounds.

The most important factor in accuracy with a Grenade Launcher is range determination. It is not possible to aim directly at targets because grenades are indirect fire weapons that fire with a high arching trajectory.

Effective use would include firing over a hill to hit targets on the other side. The grenades have a 50% kill ratio at 5 meters. Outside of that 5 meters will result in becoming a casualty from pressure or fragmentation.

Maximum Range: 400 meters
Maximum Range for an area target: 350 meters
Maximum Range for a point target: 150 meters
Arming distance: 14-28 meters

Caliber: 40 millimeter
Ammunition: Smoke, High Explosive, Cluster, and White Phosphorus

Beretta 9mm Handgun

The M-9 is a semi-automatic, magazine fed, recoil operated, single and double action pistol that fires the 9-millimeter NATO cartridge. Security Forces use the handgun when a rifle is not required to do a job, such as when the expected engagement range will not exceed 50 meters. Examples would include when posted inside a building, or on a gate shack.

In the role of personal defense, rifles inherently generate more power. Rifle rounds are also shaped differently to handle the extreme speeds they achieve. The bullet shape also aids in penetrating the target. Handgun rounds travel much slower and do not tend to over-penetrate the intended targets and continue on to hit any personnel or property that may be behind the target.

In 1985, Beretta won the manufacturer's showdown to determine which a sidearm that would replace the .38 Caliber revolvers. The M-9 is a far better handgun than the revolver in every way. Not only is if easier to operate, it also utilizes a more powerful round with a higher capacity magazine. Instead of 6 rounds in a revolver, you went to 15 rounds in a magazine that took less then 4 seconds to reload for an average operator.

The Beretta is a respected member of the Security Forces family and has the trust of the troops. It has continually proven itself to be a reliable and accurate sidearm.

During one incident on Fairchild AFB in Washington State, a Security Policeman engaged an Airman that was in the process of being discharged from the Air Force after being diagnosed as having psychological problems.

The individual brought an AK-47 on base and began shooting. The bicyde patrolman shot 4 rounds into the aggressor from more that 75 meters with two of the shots finding the target's head and ending this violent shooting spree.

Weight: 33.86 ounces unloaded and 41 ounces loaded
Length: 8.45 inches
Maximum Range: 1800 meters
Maximum Effective Range: 50 meters
Ammunition: 9 millimeter NATO

Colt M-16 Series Rifles

Since 1965, the standard combat rifle of the US military services has been the M-16; the weapon was a total new design by Eugene Stoner. Marketed as an extreme break from the Paradigms of what made weapons in the past, Mr. Stoner completely got rid of the wood stocks and wooden hand guards of most weapons prior to the M-16 in an attempt to aid in mass production capability, strength, and useful life.

High strength, molded plastics and polymers would be the new way to go. Weight of the weapon was under 7 pounds. Other weapons of the time, such as the M-14 would weigh more than 11 pounds.

The M-16A1 was implemented into the Air Force inventory to replace the M-1 and M-2 carbines. The weapon was advertised to be more accurate, lightweight, versatile, and reliable. While Army ordinance was busy trying to avoid the M-16, the USAF was the first to appropriate and utilize it.

General Curtis E. Lemay was a proponent and actually ordered 80,000 units on-the-spot at a gun show after a vender called him over and had him shoot at a watermelon in the distance. He thought it would be great for use in air base defense. Unfortunately, congress put a hold on the order, and eventually cancelled it in order to cut military spending. The USAF would have to wait for a few brief years before integration.

After initial fielding of the weapons some flaws needed to be corrected immediately. All branches of the military were reporting that the weapon had a tendency to jam up and stovepipe, leaving an empty casing that failed to be completely ejected hanging half way out of the ejection port of the weapon, not allowing a new round to be fed into the chamber from the magazine. What this leaves the shooter is an unreliable weapon. Numerous casualties would be registered due to various types of malfunctions.

Malfunctions of the first M-16's were taken care of when the Department of Defense finally listened to the Marine Corps. The Marines had run into similar problems years before during World War 2 with their experience fighting in the very humid Pacific theatre. They insured chrome was used in the chambers to avoid rusting conditions.

Later versions of the M-16 were improved considerably. They made a point to address problems that were discovered in the heat of battle. Issues such as weapons jamming, sight durability, overheating barrels, and incomplete chamber locking were all targeted.

M-16's were also given the capability to have their bolts manually pushed forward to allow the weapon to "lock" the bolt into the chamber. The forward assist sticks out of the right side of the receiver. Some eventual improvements would see the barrels were made "heavier". The benefits to a heavy barrel help to defeat overheating by creating more surface area to dissipate heat faster on the barrel. A cooler barrel maintains accuracy while it heats up by avoiding the ever so slight bending created by heat. It also helps make a barrel more durable in combat.

Other parts of the rifle were modified for better performance such as the flash suppressor. Original flash suppressors on the rifle were three pronged and succeeded in blowing gases into the ground which in turn blew dust up into the shooter's eyes. They also mud at the end of the weapon somewhat like a big fork.

Later suppressors were blocked at the bottom, which aided in keeping the muzzle from riding up during rapid firing. This was done in direct benefit to shooter control. In effect, the flash suppressor became a compensator; gases were directed up, pushing the barrel down so the next target could be engaged faster and more accurately.

M-16 M4

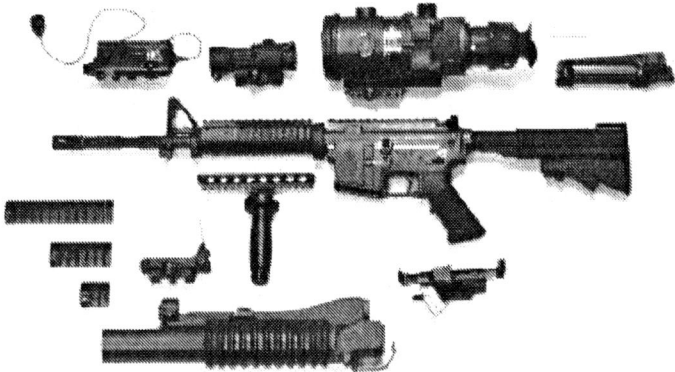

The M-16 M-4's are coming soon. The M-4's are the evolution of the shortened rifles the Air Force commissioned[3] Colt for during the mid 1960's that the special operations community made famous. This latest version of the US service weapon is perhaps one of the final upgrades in the M-16 series. Its versatility is already the stuff legends are made of. Every single Security Forces rifle need can be met with the M-4's.

There is no downside with this evolution. Troops get less weight and size while gaining more accuracy from the optical sight. Less

---

[3] The United States Air Force commissioned the shortened barrel and telescoping stock for use with the Military Working Dog program and air crew use in Vietnam

room will be used up in armories and vehicles, due to the more compact platform. Numerous additions can be mounted on a more efficient rail system. Everything from laser sights, up to night vision scopes can be fitted. M-4's are currently replacing older M-16's.

One of the concerns and issues that plagued the earlier M-16 Carbines was the lack of accuracy that came inherent with the shortened barrel. A change of bullet weight and a slight change in barrel twist ratio (the lands and grooves in the barrel) has served to stabilize the bullet better at the same time making the weapon more accurate, and thus, has served to make the rifle more lethal.

The M-16 family of firearms has been nothing but a successful venture for the U.S. military. The rifle has repeatedly proven to be a very successful weapons platform for the individual troop in the field from Vietnam to Somalia. The M-16 is easily recognizable as American and will be the primary rifle for the US military for the foreseeable future.

M-16A2 Stats
Weight: 7.5lbs unloaded, and 8.9lbs loaded (5.9 Carbine)
Length: 39" (29.8 fully retracted Carbine)
Maximum Range: 3600 meters
Maximum Effective Range 550 meters point target/800 meters area target
Caliber: 5.56 millimeter / .223 caliber
Maximum Rates of Fire:
> **Sustained: 12-15 rounds per minute**
> **Semi-auto: 45 rounds per minute**
> **Burst fire: 90 rounds per minute**
> **Cyclic: 700-800 rounds per minute**
Ammunition: Blank, Tracer, Armor Piercing

Colt M-203 Grenade Launcher

An M-203 Grenade Launcher is a light weight, single shot, breach loaded, manually operated, pump action 40 millimeter grenade launcher that is affixed to the M-16 series of rifles. In combat, the grenade launcher is used to cover dead space, or the area that regular trajectories from rifle fire cannot hit because of obstacles downrange such as berms, hills, or ravines.

The 40 mm rounds are breach loaded, meaning the entire barrel slides forward allowing the grenade to be loaded from the rear. When attached to an M-16A2 rifle, the standard blade sight that is affixed normally can be augmented with an additional quadrant sight to give more accurate engagement of targets.

M-203 Grenade launchers are a staple throughout the Air Force. Every unit that has any priority resources or supports any deployments will have these on duty or readily available at any given moment.

Grenades are effective against lightly armored vehicles and groups of individuals in a small area. Grenades come in various types from Flares and Smoke up to High Explosive Dual Purpose. The basic load for a grenadier within the Security Forces is 18 rounds of High Explosive (HE) grenades, each round having a kill radius of 50% at 5 meters. Outside of that 5 meters, fragments and debris will still cause casualties.

Weight: 3.1lbs unloaded and 3.5 lbs loaded
Length: 15 5/16" total with a 12" barrel
Maximum Range: 400 meters
Maximum Range for an area target: 350 meters
Maximum Range for a point target: 150 meters
Arming distance: 14-28 meters
Caliber: 40 millimeter
Ammunition: Smoke, High Explosive

M-249 Squad Automatic Weapon

The M-249 Squad Automatic Weapon (SAW) is a belt fed light Machine gun that fires the 5.56mm (2.23 cal.) round. Prior to the introduction of the SAW, a squad of 12 Security Police would assign one or two individual riflemen to fire bursts of automatic at any possible targets.

Even with the early M-16 that fired fully automatic as opposed to 3-round burst as does the current M-16A2, the automatic rifleman was required to fire at a rate that made him pretty much ineffective and uncontrollable with his weapons platform. He would be forced to continually change magazines that his M-16 could clear in less than five seconds under good conditions.

The SAW replaced the M-16 assigned to firing automatic or burst, support fire for the squad, hence the name Squad Automatic Weapon. With this weapon, and the same caliber round as the M-16, cover and maneuver fire can be kept up without the need for constant reloading. Keeping the round the same also provides the additional benefit of

distribution of ammo to M-16's if the need be. Versatility was also built into the weapon by allowing any NATO magazine to be loaded into the side of the rifle. If the SAW gunner ran out of belted ammo, he could load and fire M-16 magazines. Vice versa, if the riflemen run out of ammo, the SAW gunner can re-supply them. The SAW with a 200 round assault pack is a very effective and reliable platform for deterring enemy forces from advancing on a defensive position, or as a very competent ambush weapon.

Weight: 15.5 lbs
Length: 41 inches
Maximum Range: 3600 meters
Maximum Effective Range 600 meters point and 800 meters area
Caliber: 5.56 millimeter NATO / .223 Caliber
Rates of fire:
  Sustained: 50 rounds per minute with 3-5 round bursts
  Rapid: 50 rounds per minute with 8-10 round bursts
  Cyclic: 725 rounds per minute (barrel change / 200 rounds)
Ammunition: Tracer, Armor Piercing, Blank

M-60 Machine Gun

The M-60 General Purpose Machine Gun is still in use to this day. It is a belt fed, gas operated fully automatic machine gun with fixed headspace and fires a 7.62mm (.308 cal.) bullet from the open bolt position.

After World War 2, the Allies went through the defeated German military's stockpile of equipment. Everything from rations and vehicles; records to aircraft were inspected. One of the machine guns found was the MG-42 (Machine Gun design 1942). Out of this design came the foundation of the M-60.

Security Forces use machine guns as defensive weapons since that is their primary mission. A two-man crew operates an M-60. It is necessary to have the second individual as the Assistant Gunner. The assistant gunner carries extra ammo and the spare barrel kit, which includes the spare barrel, weapons maintenance tools, cleaning equipment, and a thick asbestos barrel change glove. After excessive firing the barrel will overheat, glow up, and then begin to melt down so it is important to change barrel before it heats up too much and degrades accuracy and damages the weapons destructive effectiveness.

The two key duties for a machine gun besides directly firing at a target are "grazing fire" and "indirect fire". Grazing fire is used on a Final Protective Line (FPL). The FPL is defined as the applied weapons fire in which the trajectory of the rounds does not rise above 1.5 meters, or the height of the average standing man.

Imagine looking down a flat field and shooting a machine gun in a manner in which anyone who tries to past that line would take a hit somewhere on their body, whether in the legs or the head. The M-60 allows grazing fire out to 600 meters. This is actually a "last line of defense" from a defenders standpoint. If you need to use this, you are in a bad situation.

Indirect fire is the ability for a weapon to hit targets without the shooter firing directly at them. Just as a football must be thrown upwards to fly longer distances, so must a bullet. The round may climb 100 meters in the air in route to target. In contrast, a rifleman looks though his sight and sees a target as he pulls the trigger. An M-60 can be mounted on a M-122 Traversing and Elevating (T&E) tripod.

T&E's allow targets to be scouted out and engaged at will. Whether it's a road intersection, tree line, or an old house, anything can be targeted and hit under any condition. This includes when the targets are not visible by the shooter, such as when there is low visibility or when the target is over a hill. To make this more effective, you can add an observation post that can direct fire and give corrections as required. The M-60 crew just adjusts the tripod, and pulls the trigger.

The M-60 has been in service for a very long time and has served its purpose well. Unfortunately, it is time to retire this weapon from service. Most of the M-60's have been repaired far too much over the years. It is likely that if the machine guns were needed, many would not function reliably. During training, many of the guns do not successfully feed, cycle, and shoot with the reliability that should be expected from a last defense weapon.

Although it is a favorite to many, the big, heavy, hungry, "Pig" is definitely on its way out. After years of successful and distinguished use, the Security Forces are finally supporting its removal from the arsenal.

Weight: 23 lbs
Length: 43"
Range: Max adjustable (based on sights) 1100 meters
Maximum range: 3725 meters
Maximum effective range: is as far as the gunner can accurately adjust and engage the target
Rates of Fire:
 Sustained: 100 rounds per minute (barrel change / 10 min)
 Rapid: 200 rounds per minute (barrel change / 2 min)
 Cyclic: 550 rounds per minute (barrel change every min)
Ammunition: Tracer, Blank, Dummy, Armor Piercing, Incendiary

90

M-240B Machine Gun

The M-240B General Purpose Machine Gun is the scheduled replacement for the M-60 within the Air Force. The Army contracted the design of the '240 as a direct replacement for their M-60's. Much of the stats of the weapons are the same. Both fire the same rounds, with very similar ranges.

The M-240 is just as versatile as the M-60 has been and is currently being used by the Army and Marine Corps. It will serve to fill the same roles as its predecessor and last much longer at the same time.

Performance and history has already proven to be phenomenal with this weapon. Reliability will be greatly improved and confidence from Security Forces machine gunners will be gained. No longer does the M-60 have the trust of the people who carry it. Someone listened to the calls for replacement and the only question now is, how long will it take to integrate.

Weight: 22 lbs.
Length: 49 inches
Rate of fire: 700-1000 rounds per min
Rounds: 7.62mm Armor piercing, tracer, dummy, blank
Range: Max adjustable (based on sights) 1100 meters
Maximum range: 3725 meters
Maximum effective range: is as far as the gunner can accurately adjust and engage the target

MK 19 Grenade Machine Gun

The MK-19 (nicknamed Mark 19) is a crew served, 40 mm grenade launcher. It is an air-cooled, disintegrating metallic-link, belt fed, blowback operated, fully automatic machine gun. It is capable of accurately applying a barrage of grenades to be placed on targets out to 2,212 meters for area targets.

This weapon will do severe damage to anything and everything it touches this side of a tank. Formations of trucks and troops stand no chance. With a maximum effective range of 1,500 meters for a point target such as a building, anything that poses a threat can, and will be destroyed with little problem. As a crew served weapon mounted to a tripod with a traversing and elevating mechanism, you will also be able to target areas or key terrain to engage on command, under any circumstance and sight unseen.

The MK19, 40mm grenade machine gun

was developed and tested by the US Navy in 1963 to add more firepower to smaller ships that traveled closer to land such as PT Boats, and in support of the forces whose mission it is to safely make a waterborne landing such as the Recon Marines and Navy Seals. After numerous modifications, the other military services began using the weapons. The effectiveness of this weapon has become legendary. Everywhere the platform has been deployed and utilized; it has proved calamitous to the opposition.

The MK-19 fires a wide variety of 40mm grenades. In combat roles the primary round is the High Explosive Dual Purpose (HEDP). This 40mm grenade will pierce armor up to 2 inches thick, and can throw fragments that will disable or kill equipment or personnel within 5 meters and will severely wound personnel within 15 meters of the point of impact.

Despite the modifications made over the years. There is an inherent danger of operating the weapon. Most notably is the fact that the rear end of the grenade casing acts as the chamber when the round goes off. There is spacing and a timing that must be within close tolerance.

If all is not perfectly adjusted, the gases will escape out the rear of the weapon where the shooter is. This would obviously make for a very bad day for all personnel involved. The

weapon would shoot parts of the weapon and gases to and through the gunner, ammo runner, and the assistant gunner who is spotting targets and feeding grenades into the weapon.

Firing the weapon is a blast. Loading and sitting behind such a large and heavy weapons system makes one feel somewhat omnipotent. After applying the dual thumb trigger to let a burst of 4 or so rounds go you realize how loud the gun is. You can then count 4 to 5 seconds while watching the low velocity grenades fly until they impact downrange. The gun is on target so you go for three more bursts of 6 rounds. You then stop, and watch the smoke and dust envelope the impact areas downrange. You spit the dust and debris from your mouth, wait for your environment to settle, and take note that the targets are rendered useless or non-existent.

Length: 43.1 inches
Weight: 72.5 pounds Cradle (MK64 Mod 5): 21.0 pounds
Tripod: 44.0 pounds Total: 137.5 pounds
Bore diameter: 40mm
Maximum range: 2200 meters
Maximum effective range: 1600 meters
Rates of fire:
Cyclic: 325-375 rounds per minute
Rapid: 60 rounds per minute
Sustained: 40 rounds per minute

M-2 50 Cal. Machine Gun

The Browning 50 Cal. machine gun is the father of all American machine guns. The M-2 "Big Fifty" has been around since the early 1930's and used extensively. 50 Cal. rounds destroy whatever target they hit. The round measures an inch and a half and carries extremely large amounts of kinetic energy to unlucky recipients downrange.

Targets that would be choice for the .50 Caliber machine gun include armored vehicles, low-flying aircraft, convoys, buildings, as well as equipment. Equipment could include anything from supplies to radios. According to Geneva Convention mandate, .50 Cal. weapons are not anti-personnel weapons and should not be directed towards personnel. Troops with backpack radios strapped to their backs are unlucky but legal targets however.

The gun was originally developed as an aircraft gun for fighter planes. After testing, it was then mounted to Bomber aircraft as part of their protection suites. The M-2 is utilized today as a crew served weapon system or mounted on vehicles. Many are found mounted on Humvees for Security Forces use.

When used properly in the role of base defense, the largest targets should be thinned out farther away from base with heavier weapons such as the M-2. Hostile vehicles and helicopters should be engaged with these weapons so that the lightly armed defensive posts are not forced to engage an onslaught of forces they have little capability to defend against.

Weight: 109 lbs
Maximum range: 6,800 meters
Maximum effective range: 1,500 meters
Maximum effective range for area target: as far as the
gunner can accurately adjust and engage targets
Rates of fire: cyclic, 450-600 rounds per minute
Caliber: .50 (12.7 mm)

Grenades

Hand grenades are amongst the single most powerful explosives that the average Security Forces member gets trained on. It is not like on television, the thrower feels effects of the blast throughout his body. The main issue with grenades is that the grenadier determines how far he is from the detonation by the distance he can throw.

No portrayal you can watch will ever give you the true representation of watching a grenade fly from your hands before you take cover and wait for the explosion. The explosion rattles through the entire body as the pressure passes. Security Forces utilize various types of grenades for differing roles. From combat "frag" (fragmentation), to tear gas (CS/CN), smoke, Ground Burst Simulator (GBS) and concussion.

Although fragmentation grenades may go with Security Forces when they deploy as part of logistical supply as the primary combat hand grenade for the US Military, the most common grenade is the Ground Burst Simulator (GBS). GBS's do qualify as grenades and are used primarily to train personnel, and simulate incoming ground firebombs or explosions.

The purpose of the grenade in combat is to cover area that you cannot effectively hit with any other type of firearm. Grenades destroy the targets that cannot be destroyed by guns. Although they are primarily effective against troops and are considered anti-personnel weapons, grenades carry enough punch to render light-armored vehicles useless.

Pressures can do sever damage to personnel when released inside facilities. When detonated within an enclosed area such as a bunker, armory, or command posts. Injuries such as blown eardrums and collapsed lungs should be expected.

In the unlikely event extremely high value resources (WMD)'s are stolen and there is a desperate need to recapture them, grenades will likely be thrown into their proximity as a harassment technique to prepare the area for troops to go in and regain immediate control.

On the battlefield, grenades throw

fragments 360 degrees in a 3 dimensional pattern; everywhere. They are obviously very indiscriminate and unless complete attention is paid, there will most-likely be wounded friendly forces, as well as foe.

With a casualty range of more than 15 meters (45 feet) and a fuse of 4 seconds, give or take, the grenadier must always get that grenade away from himself quickly. Grenades are trouble and should only be considered as a last resort weapon. For Security Forces operations, if grenades are used, something must have gone completely and utterly wrong, especially recognizing that the average man can throw a grenade around 40 feet.

Claymore Mine

The ultimate fragmentation grenade, the M-18 Claymore mine, is a directional fragmentation, anti-personnel weapon. The

Claymore Mine was developed during the Korean War to counteract massive infantry charges by the Chinese. The mine is made up of a curved plate that contains 700 ball bearings packed in front of a sheet of plastic explosive. Directional fragmentation mines are affixed onto trees or planted on the ground with the convex side facing the enemy.

When detonated, the mine's 700 steel ball bearings are projected as efficiently shaped fragmentation by the 1-1/2 pound layer of plastic explosive. After being initiated by a blasting cap, the explosion in turn, sends the ball bearings screaming through the air.

Security Forces utilize Claymore's in a command detonation mode, meaning that the detonation is caused by an individual who manually chooses the moment when the mine goes off once there is permission granted (with the exception of emergency) by his command element. A bit of creativity will allow a user to "daisy chain" multiple Claymores together and/or detonate them by trip wire (booby trap).

The M18A1 mine is primarily a defensive weapon. It may be employed to a limited extent in certain phases of offensive operations. The M18A1 has the same basic capabilities as other antipersonnel mines from around the world and can be used in most situations where other types of mines are used. Security Forces would set them up in the likely avenues of approach when defending a base. Defenders would detect the enemy, wait for them to come within the Claymore kill-zone and then detonate the mine.

The Claymore has an inch-and-a-half-thick curved fiberglass case. The device is six inches high and nearly a foot long. Folding steel legs can be stuck into the ground with the infamous idiot proof, "Front towards Enemy", emblazoned across the front. A blasting cap goes into one of two holes on top of the case. A wire is connected between the cap and a hand detonator. Three squeezes of the electric, M-57 firing device and the mine explodes destroying all targets within range. A Claymore can also be detonated in a non-electric fashion with an alternative blasting cap detonating cord (detcord), and a match.

The front face containing the steel ball bearings is designed to produce a fan-shaped spray, which can be aimed at a designated target area. The Claymore is an effective weapon that can be used with devastating effect in the defensive role. The $27.45 price of an individual mine makes it even more of a great value.

Range: 250 meters
Explosive Weight: 1.5 lbs.
Mine Weight 3.5 lbs.
Length: 230mm
Height: 90mm
Casing: Plastic and Green or Sand colored
Fusing: M-6, L-5A1, numerous improvised
Danger space: 16 meters, sides and rear

M-72A2 Light Anti-armor Weapon

The LAW Rocket launcher is a stand-alone, single shot, throw away, rocket system used to attack and counter the armor likely to be encountered, whether facility or vehicle. The M-72 system is a 66mm adaptation of the bazookas used during WW2.

First utilized in 1942 during the US invasion of French North Africa, Bazookas worked by allowing for the projectiles escaping gasses to exit the rear of the weapon, counteracting the recoil of the round (recoilless rifle). The LAW utilizes a rocket-powered projectile.

Early LAW rockets suffered from accuracy problems, which became severely evident once the first models hit Vietnam. Military (as well as civilian) rocket systems require two conditions to be effective; consistency and power. In order for troops to be able to aim a rocket system and hit a target, they need practice. That practice must be similar from firing one rocket to the next rocket. Taking this into account, the weapon must fire very similar each time for a troop to gain effective practice. From the start there were accuracy problems based on the stabilizer system on the rocket itself as well as the general mass production variations of the rockets.

Loose tolerance power from the rocket motors insured more accuracy problems. Power from the rockets motors varied slightly from rocket to rocket. For a rocket to be at its most effective state, an unguided rocket must accelerate to or as close to its maximum terminal velocity by the time that the rocket exits the launcher. This is less similar to a guided missile system like the Stinger missile, and more similar to a bullet as it leaves the barrel of a rifle.

LAW rockets are the mainstay ability for Security Forces to have direct impact on stopping armored vehicles. They are effective, but are ending their capability to stop modern mechanized vehicles as armor becomes more effective. The flaw with the platform is that they are unguided. Need is still there for a rocketeer to carry more than one. It may take up to three rounds of a trained individual to stop a small

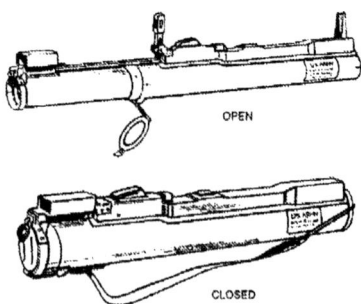

armored personnel carrier. The number of LAW rounds in possession by a Security Forces unit would likely be too few to stop a menial, mechanized attack. Positions would likely be overrun as a unfortunate conclusion.

Accuracy could not be gained without successfully ensuring all rockets going to the field were as close to identical as possible. Once the power differences were addressed, a better product could be produced.

The latest generation of LAW rockets evolved into a much better system. Many of the mass production and quality control problems were addressed and this weapon was still in production after more than 3 decades of use. A more effective, and versatile replacement is in order.

AT-4

As the Army's primary anti-tank weapon, the AT-4 is an 84mm, high explosive anti-armor, fin stabilized, disposable launch tube rocket system. In comparison with the LAW, the AT-4 fires a more powerful round with greater armor piercing capability.

Although the Maximum Effective range is 250 meters, the round will fly out to 2100 meters. It weighs just under 15 pounds. There is no need to extend the launch tube prior to firing and there is a capability to attach night vision equipment to the launcher tube addressing one of the main issues with the LAW. The unit price is a very economical $1480.64, which is worth the ability to stop modern armored vehicles.

The latest anti-tank rockets that the Army and Marines are employing are slowly making their way into Air Force service. The AT-4, Anti Tank missile will be a substantial increase in defensive capability for the Security Forces.

Length: 40 inches
Weight: 14.8 pounds
Caliber: 84 mm
Maximum Range: 2,100 meters
Maximum Effective Range: 300 meters
Penetration: 400 mm of rolled armor

Mortars

Mortars have been a staple of military service and were first used as early as the 14th Century. Mortars are currently the last muzzle loading heavy weapons platform in use by the ground forces of modern militaries. Because of its compact size they have a beneficial Firepower to weight ratio and can be transported with minimal support.

The United States Air Force Mortar program has been in a constant tug of war. During the Vietnam War, mortars were very beneficial and skill utilized by the base defenders was expert. After the end of the war and up to present there has been a push to utilize this "Heavy Weapon" platform as simply an illumination platform. The lessons the Air Force learned during the Vietnam War concerning the benefits of the mortar are being quickly lost.

As the platform with the greatest range of all Security Forces heavy weapons, mortars can reach out and touch the enemy, however the Security Forces leadership has failed to push Air Force leadership to understand the capability, destructive power, and base defense potential the mortar possesses.

Training still takes place with mortars in Indian Springs, Nevada by the 99th Security Forces Group; however, Security Forces mortar technology has not been upgraded to the level it should be at. While the Army and Marine Corps understand the value of the mortar and have maintained and advanced their programs, the Air Force has not done the mortar, or base defense, justice.

Security forces are still relegated to manual means of plotting targets as opposed to some of the newer computerized and laser targeting mortar fire control methods. Accuracy is not a problem for Mortar crews in the Air Force, but time to manually plot targets could be better. Technology and training should definitely be addressed and improved.

For such a distinguished weapons platform that has been used in its current form for more than 80 years and has proved it's lethality. To go under utilized is not appropriate. It is not utilized as effectively as its potential dictates. As a minimum, more time spent teaching forward observation and call for fire methods would benefit base security. Air Force mortar support, as well as Army/Marine artillery pieces that may be in the support range could then be utilized.

Security Forces train as if they have artillery support, but reality says that defenders would not be able to utilize the resource due to lack of communications between services. Mortars are the most neglected weapons platform in the Security Forces inventory, as well as the most powerful.

The 60mm and the 81mm mortars in use with the Security Forces are very effective when used correctly, the 60mm reaches out to a maximum distance of 3,500 meters, while the 81mm mortar can reach out to 5,800 meters. If necessary, a ground force commander can destroy targets before they are standing outside of his base perimeter threatening air ops.

Stinger Missile System

The Stinger missile system is designed to give ground forces the ability to defend themselves against aircraft attack. The Stinger is somewhat effective against fast moving aircraft, but where the Stinger missile is most effective is against larger and slower cargo aircraft and Helicopters.

The Stinger is a man-portable 1.5 meter long, 34.5 pound, shoulder-fired anti-aircraft missile. It is designed to counter slower, low flying ground attack aircraft. The missile is guided to its target by an infrared, heat seeking guidance system. The Stinger's "fire-and-forget" homing ability allows gunners to take cover (run and hide) or to engage new targets with another Stinger immediately after firing one missile.

Packed with 2 pounds of explosives, Stingers can do severe damage and pose and extreme threat to enemy aircraft operations. The greatest example of the Stingers capability combined with minimally effective training and use is the Soviet Union's Afghanistan conflict. The Afghan Mujahadeen brought down more than 270 assorted Soviet aircraft in their fight with Stinger missile systems. That equates to roughly about 79 percent of aircraft that were engaged with the Stinger. Numerous other aircraft were hit, sustaining damage, but were still able to limp home. This weapon proved extremely successful.

The current version of the Stinger is used in conjunction with the IFF system. IFF refers to "Identification, Friend or Foe". This is one of the safeguards that allow US Air Force aircraft to work around those aircraft of the sister services, as well as the aircraft of friendly nations and allies. The IFF signal notifies the Stinger Crew that the targeted aircraft is transmitting back information that may identify it as friendly informing the Missileer that it should not be engaged.

The main area of operation for Security Forces air defense artillery capability is the Republic of Korea. Of course, the United States Air Force is currently not using the Stinger Missile System at the level that it should be. There is a discrepancy that needs to be addressed where air attack is concerned, and that discrepancy will likely not be addressed until after some incident involving aircraft that could have been stopped through the use of a Stinger actually occurs.

Unfortunately training and funding are the main issues. Training with the stinger is significantly pricey and that must be weighed with the risk of an aircraft attack on an Airbase by an enemy aircraft. Such an attack on an airbase has not occurred since the Korean War. This is also a factor why a Korean Security Forces unit is tasked with the Stinger (1998).

With a range of one through six kilometers, based on target altitude, and a maximum flight ceiling of up to 10,000 feet, low and slow flying airplanes and helicopters are pretty much easy targets.

This $40,000 supersonic missile is also made to survive the rigors of military transportation and handling. It is very sturdy with pains taken to insure that none of its parts are fragile. Built by Hughes Missile systems with research and development by General Dynamics and Raytheon Corporation, it is obvious why this has been such a deadly effective anti-aircraft weapons platform.

Length: 5 ft
Weight:
       System: 34.5 lbs
       Missile: 22 lbs
       Warhead: 2.2 lbs
Diameter: 3 inches
Guidance: Infrared-homing
Speed: 2,300 feet/second
Range: 650-16400 feet
Warhead: Proximity fuse with penetrating high explosive

Vehicles

Long gone are the days when Security Forces stood around and "humped" aircraft. There was a day when vehicles were not as widely available as they are now. Believe it or not, most security troops stood next to Airplanes for their entire shifts. 8 or 12 hours of standing up and walking around an aircraft to do their job was not uncommon.

Walking from wingtip to wingtip counting steps and/or rivets on the aircraft earning cops the nickname "Rivet Counters". This obviously was not as fun as it sounds. Anyone can imagine how the complaints from the troops must have gone. When Security Police had a chair, they could not be more content.

It was perceived that Security Police would not use vehicles effectively since they did not have to drive anywhere. They would stick around one post until they were relieved. Not only were most security units not allocated many vehicles; they accepted the fact that they did not need any vehicles. Unit Commanders, Officers, Chiefs, along with other senior enlisted didn't ask for more vehicles because they had been raised in an Air Force where they didn't have those vehicles to do the same job; so why get them.

Eventually, the idea of a "kindler and gentler" Air Force would change perceptions. More and more vehicles would make their way into service and on to post. Whether as law enforcement patrol cars, security patrol vehicles, tactical vehicles, combat patrol vehicles or utility vehicles, any vehicle could find a home in a security squadron.

From small two door economy cars, compact mini trucks, all the way up to huge diesel F-350 6-pack, four-wheel-drive trucks; Security Forces can use any vehicle. Motorcycles have also been used for various Security Police activities over the years, most interestingly as traffic control vehicles.

The Jeep

No history of military vehicles could be complete without mention of the Jeep. Since World War 2, Jeeps have been a fixture of practically every military unit that had vehicles until the mid 1970's. Jeeps were designed to be simple, durable, and versatile. Truly designed to go anywhere at anytime.

The name Jeep came from the troops who drove and worked with them during the Second World War. In typical military fashion the original duty of the vehicle was General Purpose. You would not hear troops asking for the keys to a "General Purpose" vehicle; naturally this got shortened down to the "G. P." This GP moniker mixed with a Popeye cartoon character named the Jeep, who could go anywhere, do anything, including disappear, sealed the fate of the namesake Jeep vehicle. The debate continues to this day, but the most likely answer is a mix of both truths.

Jeeps have done everything. From the transport of troops, scouting, towing trailers, and even being a mobile weapons platform. Although there were numerous attempts to replace the many roles of the Jeep, there would be no true replacement of the Jeep until the HMMWV (Humvee) appeared in the mid 1980's. This is a true homage to the position the Jeep has held within the military.

Security Forces of the Air Force have continually used the Jeep for many roles. Jeeps have seen duty as security response teams, tactical patrol vehicles, convoy duties, and have even been up-armored for better crew protection while responding to incidents. Jeeps were in a few security units until as late as 1995. Unforgettable and fun to drive, jeeps have served their country with distinction.

Since its inception, the Jeep has been updated and modernized continually with millions of the vehicles seeing military service in peacetime and war. The versatility of the American Jeep is legendary in both the military role as well as the civilian.

2-½ and 5-Ton

The M-35A2/3 general-purpose 2-½ ton (deuce-and-a-half) trucks have been the workhorses of military units throughout the world. These trucks have been around in one form or another since pre-World War 2. Since the Air Force became a military branch, it has utilized these trucks mainly in Europe. This family of trucks has been used for everything from towing equipment like water buffaloes (500 gallon drinking water tanks) up to transporting troops out to post.

The key to the trucks use is support. They move units and equipment from place to place. Even though the 'deuce was very capable. Larger equipment would require still larger vehicles. The 2-½ ton truck would morph into the 5-ton truck. The 5-ton version would see many improvements, however the version most widely utilized of the 9-series (5-ton) would be the M-939, which was manufactured between 1984-1986.

Both variants of these general-purpose trucks share most of the same characteristics that made these vehicles so successful. Both vehicles look similar except to the extremely knowledgeable observer but it is an illusion. The fastest way to determine which version is which is by size.

There is a huge difference that one notices when in their presence. The size difference cannot be missed up close, but easily mistaken at longer distances. The biggest difference is to note the locations of the fuel tanks. 5-ton's have fuel tanks and accessory boxes farther back under the cabin doors in a more subdued position while the Deuce's sits more prominent.

Both of these truck variants are full time 6-wheel drive vehicles that are fairly competent in less than perfect conditions. The drawbacks to using these vehicles off-road are that the center of gravity is placed high in proportion, and that the weight of the trucks is desperately high. The rating of these vehicles is based on the load carrying ability of each axle; 2½ ton and 5-ton.

The 5-ton vehicle was made to replace and accomplish what the 2 ½ ton could not do. Unlike the deuce-and-a-half, the 5-ton has an automatic transmission. This definitely makes driving much easier for the less skilled and perhaps less trained driver.

Versatility was built into these trucks. Designers recognized that they may not always have access to "high quality" American diesel fuel, so they were built with Variable Fuel Engines (VFE). A rule of thumb is that if a vehicle can run on a fuel, you can stick it into these trucks. From diesel, to unleaded, up to jet fuel, the engines can burn and operate off of them. Wherever you are, you should be able to find some kind of fuel to keep you pressing forward.

Survivability has been a key to their longevity. These vehicles are solid and built to last. As homage to the vehicles is that not only have they been around a long time, they have went though a modernization process. Old vehicles are taken apart, stripped down and rebuilt with updated parts, equipment and systems such as driver controlled tire pressure valves.

These vehicles are amongst the most recognizable vehicles in the American military's history. Wherever you go in the world that the United States ever fought or deployed to, the people will recognize them as American trucks.

Cadillac Gage Peacekeeper

  In the early 1970's, the United States Air Force required a versatile vehicle that could be standardized by Security Police units worldwide. In 1976, the Air Force brought about the Peacekeeper. All in all, the Peacekeeper project was successful in placing a lightly armored vehicle that could provide some protection to personnel responding to most incidents. The vehicle had to support everything from normal day-to-day flightline driving, fire team operations, air base ground defense, as well as missile field convoy and patrol operations.

  Driving a Peacekeeper (PK) was always a daunting task that required constant attention. Visibility was so poor one might as well have been driving a tank. The weight of the vehicle was astronomical at over 4 tons. That is, more than 4 tons of metal built on a Dodge 1 ton truck/van frame. With that said, the brakes were very underpowered for a vehicle of its weight. If you were forced to make a turn too fast you clenched your teeth, puckered up, rode it out, and prayed everything went OK.

Peacekeeper center of gravity was so high because of all of its top-heavy weight in armor. Distribution of that weight made a person not used to the sensation of turning feel like the next turn could be his last. In actuality, one never really got used to turning. PK's were very dangerous because of this condition and numerous rollover accidents resulted from their use.

The suspension needed to be so stiff to support the weight that when combined with the solid rubber (bullet proof) tires, the crew's guts took a beating over any and every bump or crack in the road. The size of the vehicle wore the springs and shocks out pretty quickly. They were simply, not up to the task assigned to them. Most drivers used to pray for a PK with fairly worn out springs, however. Turning with worn springs was worse, but the straight line driving would be less obnoxious over the bumps, and road cracks. Worn springs allowed you to kind of, bounce down the road.

Although a successful project in general, it was short sighted and could have been much better. But it was notable that security leadership, actually stood up, talked to the Air Force leadership and made this project happen. This does not happen as often as it needs to.

The PK's had a life lasting over 20 years of use. Security Forces no longer utilize them. Peacekeepers can still be seen in the service of civilian police departments throughout the country. Police organizations were offered these vehicles as they retired from service.

The replacement for this armored vehicle requirement is the "Up-Armored" Humvee. Specifications were given by the Air Force and a vehicle was designed; yet the Air Force has failed to get the Hummers out to the field at a significant enough rate to replace the Peacekeeper.

AM General Humvee

The AM General High Mobility Multi Wheeled Vehicle (HMMWV) came about in the early 80's when military funding was great due to the Reagan administration's drive to rebuild the US military. The idea was to provide a highly mobile vehicle system that could cut down on the number of differing types of automobiles required in the military.

Instead of a vehicle type for tactical engagements, another for ground reconnaissance, and yet another for patrolling, there had to be a way that one vehicle could cover all these differing requirements; the hope was to increase capability while reducing vehicle fleets.

You need a spectacular vehicle to replace the tradition and versatility of the amazing Jeep. The replacement must be so capable it makes us all wonder why we thought the Jeeps were so good. The Jeeps were great vehicles that could go anywhere and carry or tow a load. Problem was that you had to be an extremely skilled driver to exploit the higher capabilities of the Jeep and your load was going to be pretty miniscule.

Humvees on the other hand can go anywhere with an average driver. Not only can you go anywhere, you can also take more weight, with more personnel. Humvees can be mounted with any array of weaponry from light machine guns all the way up to anti-tank or anti-aircraft missile systems.

The last promise of the Humvee is the armored version. The "up-armored" version was slated to replace the aging fleet of

Peacekeepers and finally allow them to be removed from service. The problem however is that the vehicle was not being bought at an adequate rate to allow the timely retirement of Peacekeeper. Currently throughout the Air Force, the Peacekeepers are long gone with no up-armored replacement. This should not have been at issue since the Air Force was the main branch requesting the fully armored package variant.

Cadillac Gage Commando

One of the more interesting vehicles used by Security personnel within the Air Force for air base defense is the M-706, Cadillac Gage Commando. The Commandos were made to provide some real protection to Security Police during actions of the Vietnam War. The Commando was massive in size with almost 16 inches of ground clearance and tires that were more than 4 feet tall. Average operating weight sometimes exceeded seven tons.

121

The '706 was designed for convoy escorts, armored response, reconnaissance, and police riot control. Production began in April 1964. A large number of these vehicles saw use in Vietnam by not only the Air Force, but also the Army and Marine Corps. Many platforms of the vehicle were adapted for specific uses. Security Police armed the vehicles with twin (forward and rearward facing) M-60's, twin .50 cal. machine guns, as well as twin mortar crews.

Rifles used by the crew could also be fired from the various portholes located around the vehicle. The Army and Marines had variants armed with even larger weapons from their .50 (12.7mm) caliber machine guns up through to 40-millimeter cannons.

This was a truly capable and versatile vehicle. It was also a true amphibious vehicle. Not quite a duck, but it did qualify to operate on water, as long as the drain plugs on bottom were completely secured insuring that the vehicle would float. Waterborne motion was provided by the tire treads "swimming" through the water. When you couldn't get out of the water, the front mounted winch could always help pull you out.

With a hull of welded, high hardness ballistic armor plating, crews were more than protected against small arms fire, grenades, and anti-personnel mines. Personnel felt safe, albeit a bit cramped on the inside. The crews could include up to twelve troops with all their associated gear and weaponry.

These vehicles were gained through the Army's supply system, which slowed their requirement for the M-706 after the contract had been secured and production was under way. Base security benefited in Vietnam with more than 60 of these vehicles.

WIDTH: 7 ft. 5 in.
LENGTH: 18 ft. 6 in.
HEIGHT: 6 ft. 4 in.
WEIGHT: 16,250 lbs. loaded
ENGINE: One Chrysler 361 cu. in. V-8 gasoline engine of 200 hp.
POWERTRAIN: Manual 5-speed transmission with 4-wheel drive
CREW: 12 including driver

Maximum Speed Land: 62 mph.
Water: 3.4 mph.
Range - Road: 550 miles
  Cross Country: 400 miles

M-113 Armored Personnel Carrier

The M-113 is a fully tracked, armored personnel carrier (APC). It provides protection to Security Forces as they move into a possible combat area. Along with protection, the M-113 increases mobility for personnel and cargo. As a fully armored vehicle weighing 27,200 pounds, it can carry up to 11 troops, the driver and co-pilot to and through most terrains.

The Air Force M-113 armored personnel carriers operated within Vietnam utilized by many larger Security Police units there. These vehicles made Security Police more capable of getting to the areas the enemy was at. Protection provided for the individual made traveling over terrain a bit more safe even though the loud and large vehicles made delicious targets for enemy forces. The M-113 is still in service with the Security Forces in a few places today, most notably Korea.

Just as is the case with most US military combat vehicles, numerous weapons platforms can be mounted topside. From M-60 machine guns up to Mk-19's and .50 cal. machine guns. One of the most effective weapons the security forces have ever utilized in its more than 50 year history was also mounted on the M-113's in the form of the M-134 Mini-gun. The Minigun is an electrically operated, rotating multi-barreled, machinegun. Having a gun with more than one barrel sidestepped the problem of barrel overheating. If you add an additional barrel to a machine gun, it can be operated twice as long. But if you add four extra barrels, you can operate the gun four times as long, and hopefully, that additional time won't be necessary.

The M-134 is a direct descendant of the Civil War guns designed by Dr. William Jordan Gatling. Miniguns had the benefit of an extremely high rate of fire, and they were jam proof. They did not rely on the capture of any gases, or springs to operate. If a round dd not fire it would be ejected just the same as if it had been fired. The Cycle of operation was only dependent on electricity.

The M-113 also goes down in history as the only tracked vehicles that Security Forces have utilized. Tracked vehicles  are great for extreme terrain; however, they do not handle well on hard surfaces such as concrete, which most air bases utilize a lot of. Security units would be limiting themselves with vehicles that would not be able to be driven competently on hardened surfaces without causing damage to the surface itself. Additionally is the possibly of causing foreign object damage (FOD) to high performance aircraft engines.

More than 30 of these vehicles were appropriated and utilized by Security Police in Vietnam and many more have been used since. These vehicles still see limited use; again most notably in the Republic of Korea.

Length 191.5 Inches
Width 105.75 inches
Height 86.5 inches
Weight: 23,880 lbs
      Combat load: 27,180 lbs.
      Max weight: 31,000 lbs
Turning radius: 0
Center clearance: 16 inches
Trench crossing: 67 inches
Cruise range: 300 Miles
Max speed: 41 Mph

Fast Attack Vehicles

Occasionally a jolt of creativity brings in resources that really do a great job of benefiting the mission of the Security within the Air Force. Edwards Air Force Base Security Police picked up 2 Fast Attack Vehicles from of all places, the Defense Reutilization and Marketing Offices (DRMO). DRMO is the organization that sells equipment that the military no longer needs. The price for the vehicles couldn't be beat at the whopping amount of "free".

These vehicles started off as the initial test vehicles to what would become the current model buggy used by the Special Forces, Navy SEALS and other Special Operating Forces worldwide. Military Sand Rails are now

armed with weapons to include Mk-19's and

.50 Caliber machine guns. Anti-tank and anti-aircraft missiles can be fitted.

"Dune Buggy's" would normally be used as recreational vehicles by weekend warriors to race throughout the Desert, however, these were used to secure the miles and miles of desert area in support of the Air Force Flight Test Center landing site for NASA's Space Shuttle program. Besides the Environmental concerns, this was another example of a great idea to use a specialized vehicle to aid security

Patrol Cars

The babies of most security forces vehicle fleets are their patrol cars. Just as in any civilian police department, much of the professionalism thrust upon them is based on how they portray themselves. Nothing shows unit pride and authority quite like a full out high performance black and white. Patrol cars utilized by Security Forces have always strived to mirror the patrol cars that civilian departments use. The problem is always funding. With the exception of the mid 70's through about to the early 80's big, four door police cars ruled the roads. Big gas guzzling cars from the late 40 to the early 60's were prevalent as the price of gas was low and the protection of the environment hadn't even been a glimmer in anyone's eye.

During and after the OPEC fuel embargo, boxy, ugly, small "cookie cutter" two door, K-cars were forced upon Security Police due to poor funding and the whole kick on fuel efficiency. It would take a few more years to see the return of the large Ford LTD's and Chevrolet Crown Victoria's.

The need was in having cars that could get better fuel economy. The unfortunate downside was a car that had no power and was cheaply manufactured. It is said that if a vehicle could withstand the Cops, it could survive anyone. Security units destroy most cars and the mid 70's cars were definitely no exception. They were destroyed fast; they just seemed to hang around for far too long because money wasn't available for replacement.

Modern Security Forces patrol cars have even more in common with their civilian counterparts. In fact, most the cars that are ordered come from the Government Service

Agency (GSA) and are exactly the same models purchased by civilian police departments across the United States. They are built with police Interceptor packages that are specifically manufactured with all of the wiring that is required to hook up the exterior lights and control boxes for your radios, flashers and overheads. The vehicles are also reprogrammed and modified to produce an increase in power as well as being fitted with a more aggressive suspension. This definitely makes sense for law enforcement patrols.

The Red Hats (CATM)

The internal defense of USAF bases and the survival of downed aircrew members may be dependent upon individual proficiency with assigned firearms. All Air Force personnel have defense responsibilities against overt and covert enemy action. To discharge these responsibilities, the fundamental military concept of competency with firearms must be reinstated within the Air Force."

--General Curtis E. LeMay

As a practical safety precaution, Air Force firearms range officials wore red hats as a means to distinguish the trainees from the instructors. The red hats became a regular occurrence and have always worked in close conjunction with Security Forces, who were usually the largest customer to AF Combat Arms Training and Maintenance (CATM) units on most bases.

Kimpo Airbase was effectively overrun by a large unit of communist Chinese forces during the Korean War. The Kimpo investigations combined with other instances where Air Force personnel · failed to use firearms in a proficient manner got the attention of General Curtis E. Lemay. General Lemay was an avid shooter and a true proponent of Air Force Security although history recognizes his ideas on Air Power alone. General Lemay saw the need for all AF personnel to be trained in marksmanship and firearms maintenance.

Curtis E. LeMay
USAF Museum Photo Archives

General Lemay championed the Small Arms Marksmanship Training Unit, or SAMTU within the Air Force with a purpose to prevent a recurrence of Kimpo Air Base's overrun by providing weapons training to Air Force personnel. To this point, more than 6500 instructors from the Air Force and its components (Air National Guard and AF Reserves) have trained to become Combat Arms instructors since the early 1960's.

SAMTU did support the Air Forces marksmanship teams to the point that competitive shooters were allowed to cross train into SAMTU. SAMTU evolved into

Combat Arms Training and Maintenance (CATM) over the years to reflect a more realistic view of their actual duties. CATM was responsible for training and qualifications of the air base populations as well as the maintenance of the entire bases' armories and weapons caches.

Although differing organizations on Air Bases are required to stay competent on firearms and marksmanship, the main customer of CATM has always been the security unit on the base. In 1982, CATM would be put under operational control of Security Police units. 1997 not only combined the Security Specialist and the Law Enforcement Specialists. It also combined the Combat Arms into the Security Forces as well.

The demise of the traditional Red Hat was included into this merger. Currently, Combat Arms personnel are trained as regular Security Forces. An application process is used to determine which Security personnel go on to specialized training to become Combat Arms Instructors.

Firearms experts also serves to add versatility to the personnel that are deployed to isolated locations by giving them personnel that can handle the most severe weapons associated issues on site, as well as fight if need be. Unfortunately, as a testament to the end of the era, red hats are not generally worn by range officials anymore.

Reserve Components

Traditionally the Air Force has maintained the closest ties to the civilian police than the other services have. The USAF policy of "Total Force" insures all Air Force resources are recognized as being vital to the mission we accomplish. Without the components of the Air National Guard and Reserves, it is unlikely that the Air Force would be as successful as it has been. From Aerial Refueling and Logistics to Special Operations, the Air Force reserve components are imperative.

Security Forces generally recognize the knowledge of its "part time" personnel. Army and Marine MP's are more closely tied to their wartime missions and many of the adaptations the Air Force attempts are all but unthinkable to a soldier, sailor, or marine. As opposed to being directly tied to direct combat roles, AF security is a daily duty. In a combat zone, there may be more posts with more personnel being utilized, but the mission remains the same. Air Force security personnel will even carry the same gear and weapons and operate generally the same as any other day.

Many of the updated policing skills can be directly contributed to the fact that reservists, Individual Mobilization Augmentees (IMA's are reservists assigned to active duty units), and Air National Guardsmen that work as civilian Police Officers have a very positive and somewhat influential position in their Air Force units.

Generally these professionals are understood and recognized to be subject matter experts in all areas pertaining to practical law enforcement since it is their trained profession. Most Commanders welcome trained police into their units and listen to what skills they have to offer while utilizing the latest law enforcement intelligence that can be shared.

Within the AF, everything from holsters to high-risk traffic stop procedures are more or less, borrowed from the people and Police departments who use these skills everyday. Law enforcement skills are very perishable. Along with being perishable, skills and tactics change routinely. What works great one year varies from what works the next year.

Civilian police departments run through scenarios as part of their jobs and are forced to alter their operational tactics in order to keep ahead of the "Bad Guys". This updated information that Reserve Security Forces provide always prove valuable to the Active Duty Air Force units they operate with.

Emergency Services Teams

Just as the case is with the SWAT programs of many Police departments throughout the United States, the Security Police followed the example of the first SWAT program, which is/was the Los Angeles Police Department. Los Angeles determined that they had a need for dedicated personnel to meet certain challenges that began to arise during the course of duty from drug and gun violence. Whether there was a need or not, Air Force units followed suit with tactical teams of their own. Initially, teams had no direction but slowly, Security Police command would begin to formalize training for all tactical teams. An Emergency Services Team (EST) program would be created Air Force Wide requiring most bases to have an EST program.

Within the Air Force, the basic purpose of an EST is to provide more practiced and motivated individuals when/if a need may arise. An EST would supplement other emergency services such as Explosive Ordinance Disposal, Medical, Fire Fighters, as well as SP Hostage Negotiators. As a last resort, an EST is capable of entering an area to bring a situation to a more controlled, less random end. The idea of a formalized course and standard was a great idea on paper, however, may have led to the premature downfall of the program as a whole.

One of the failures of the Air Force EST program was the training course was too hard. The attitude that was fostered was one of a Special Operating Force as opposed to a law enforcement agency; meeting standards of physical conditioning that were higher than Army Ranger PT test standards was unnecessary.

The school failed to produce personnel that could go back to their bases and act as cadre to teach tactics to wider groups of security troops while at their bases. The Air Force program morphed into a course that provided the only EST personnel. The course trained personnel that would eventually make up entire base teams. The course was so demanding that Commanders chose not to send personnel because the likelihood was that they would fail. If there was one single cause of the program to die, it may have been this lack of support at the Squadron Commander Level.

Personnel that were successful in completing the school didn't always learn the latest SWAT tactics of setting up on situations and killing targets. They could run extremely well and do hundreds of pushups, but that's not necessarily what an EST needs to be successful during a tactical situation.

Another failure is that the Air Force was trying too hard to force a regulation on EST standards and tactics. To create a standard that reaches across the entire Air Force and requires all teams to operate the exact same, with the exact same formation. Creating an operating regulation was not beneficial whatsoever. Natural unit competition across the USAF should have taken over allowing the best teams to shine and the worse to suffer.

Civilian SWAT and tactical teams across the United States train as they need. LAPD has been very key in team developments, but many improvements and valuable ideas have come from sources that varied from the original LAPD SWAT model. Ideas such as shoot house training, tactical use of mirrors, flash bang distraction devices all stemmed from departments outside of LAPD and have become standard practice.

If you feel the need to have a regulation, this should be a resource to set a minimum standard, not a rule that insures exacting standards of how to do everything

from personal movement tactics up through team organization. Flexibility should be left up to the unit based on their size and the needs that they posses for their localized area.

Some of the skills that practiced and supported EST teams give a Commander are patience, additional firearms training, shooter discipline, small unit tactics, and sometimes even rappelling capabilities. These skills supplement the basic skills that all Security Forces know such as use of force, and Individual tactics.

Although there have been more that 70 "Callouts" for EST's when they existed, the height of the program was in March 1979 and was initiated by the Army Provost Marshal of Fort Sam Houston in Texas. A distraught suspect barricaded himself in a room after shooting at MP's, killing one. EST cadre from

138

Lackland AFB was notified and responded.

The EST arrived and secured the inner perimeter and came up with an entry plan as the negotiators made contact. The suspect would eventually be removed after hours of negotiation. But after this, EST would begin to build a reputation based on its professionalism, training, and skill.

Ironically, although the Air Force initiated this DOD tactical team concept, the program would be doomed to history. The school would close to new students due to lack of funding and support. The US Army runs the current DOD tactical course.

Emergency Services Teams are dead. They are a successful moment in the history of the Security Forces. The current push is for "Tactical Teams". There is no reason for a unit with more than 150 troops to fail to provide 15-20 people who are extremely competent, motivated, practiced volunteers to stay trained in tactical operations in case the need arise.

Military Working Dogs

The Military Working Dog (MWD) has proved invaluable to the USAF Security Forces. Specifically trained dogs have been used by a vast number of military organizations. In every role performed from attack, hunting, searching, as well as their universally recognized capabilities of personal protection. Canines have truly been a Security units' best friend.

Security Forces detection capabilities were drastically increased instantly upon the MWD's introduction. The dog enhances detection capability in order to protect the resource whatever that may be. Obviously not a new technology, the Security Police took the lead in training and doctrine in the use of the Dog. The Air Force would eventually lead the Department of Defense in providing this capability by training, not only military dogs, but also other federal, and even, civilian Police dogs.

In the 1950's, the Army made it clear that it wanted to do away with its entire canine program. They felt that their need for dogs had diminished. To a certain extent the Army was correct. The Army didn't really have a large need for sentry dogs. The Army had quite a different outlook and requirement than the Air Force as far as security was concerned.

Traditionally, the majority of Army posts/bases were not secured in the manner that air bases were, many were even military "reservations", open for general civilian traffic. Army bases can be extremely large ranges placed in remote areas that would make it impossible to fence up and secure entirely.

The average resources on Army bases were not usually big-ticket items whose loss was detrimental to the mission of the Army. When resources required real security, those particular areas of the base were fenced in, locked down, if required, personnel were posted. Examples of this were armored vehicle lots, Armories, and Intelligence facilities. The majority of Air bases on the other hand, were secured everywhere. Fences were placed to insure only authorized personnel were present from the base's perimeter on.

During Vietnam, Security Police made a change of direction in Military Working Dog use. Dog were no longer simply sentry guards. The new emphasis moved to patrol dogs that could be deployed outside of the fence lines in order to detect the enemy as far away from the base resources as possible. One such engagement by working dogs showed the extent and value of the canine in the role of security. This engagement took place in the early morning of December 4th, 1966 at Tan Son Nhut Air base, Republic of Korea.

The Base was attacked by a 60-man force of Viet Cong commando raiders who used a single avenue of approach through several security posts. Once inside the base, the raiders split into small groups to go after various resources. Several dog teams alerted practically at the same time.

Immediately, Rebel, a sentry dog on patrol, was released. The response was a hail of bullets that killed the dog. Forty-five minutes later, the group was detected by another dog. Cubby was also released only to be shot and killed as well. Yet another dog, Toby, was killed and several handlers were wounded. As a result of this early warning, Security Forces of the 377th Air Police Squadron quickly repelled half of the attack and prevented damage to aircraft, personnel, and facilities.

Another dog team was wounded but followed the enemy long enough to give information to other posts and have a machine gun bunker set up in ambush. As the raiders made way to the aircraft-parking ramp, the bunker opened fire and killed that entire attacking group. As more Security Police made their way to the base perimeter, they were able to prevent the majority of infiltrators from escaping and forced some remainders to hide within the base perimeter.

Without, MWD's, search squads found no other enemy and called off their searches. Unfortunately, they did not completely search a graveyard, some dense vegetation, and some wells. As the following night approached and

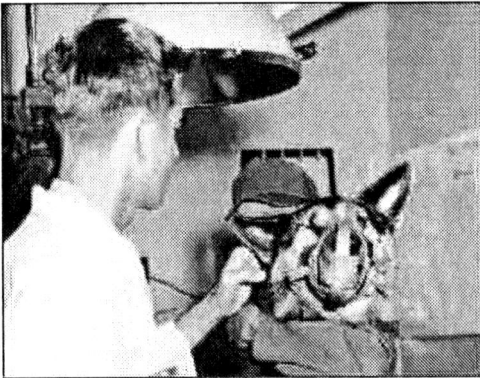

the dogs came back on duty, A2C (Airman Second Class) Robert Thorneburg and patrol dog Nemo were out on patrol when Nemo alerted on something and was released to attack only to be opened fire upon.

Both Thorneburg and Nemo were wounded in this engagement with the remaining VC, but only after killing one more VC. Nemo's injuries included the loss of one of his eyes. Eight more VC were discovered hiding on subsequent searches and were killed. Nemo was credited with saving Thornburg's life as well as preventing further loss to the USAF. He was later retired to Lackland Air Force Base.

The use of Military Working Dogs throughout the DOD is on the increase once again. The Army and the Marine Corps are starting to show regained interest in Dog use. This is in response to the detection capability and the increased need of security on all installations. Some of this can be directly attributed to the current war on terrorism. Within the Air Force however, the value of dogs has never been questioned.

As the Air Force becomes continually more successful in its expeditionary role, MWD's will become exponentially more valuable to security of the numerous smaller sites that the Air Force will need to operate from. Bomb sniffing dogs are getting a workout at the entry control points of many locations currently and for the foreseeable future.

The benefits of having sentry dogs around nuclear resources cannot be disputed as an additional detection capability. There should not be a decrease in Military Working Dog use for a very long time. MWD's are also partially responsible for the decrease in Drug use (or increase in detection) Air Force wide from the mid-1980's alongside of the negative view of drug use and the random drug testing programs mandated by the USAF.

From detection of intruders, explosive detection, and even down to quelling fights at the base nightclubs. Whenever dogs show up, Security Forces are instantly more effective.

The Elite Guard

In 1956, Strategic Air Command (SAC), at the direction of General Lemay, would once again drastically raise the standards for Air Police. Lemay would order that an elite detachment of guards be handpicked from around the Command, and posted to secure the Headquarters as well as provide personal protection for the Commander and Vice Commander of SAC. At the time, he was Commander in Chief, Strategic Air Command.

The following year, in 1957, General Curtis Lemay would hand select the new look for the Elite Guard uniform. The uniform would consist of khaki slacks and shirt, jump boots were worn with the slacks bloused into them. White scarves (ascots) would complement white bootlaces and the blue beret with the SAC patch affixed.

146

All later Elite Guard uniforms would be modernized to reflect newer shades and fashions of the Air Force uniform of the day. It is of no doubt that all uniforms were just as impressive looking in their respective eras. Representing professionalism, the Elite Guard won respect and accolade worldwide through their professionalism and performance.

Although SAC is gone due to AF restructuring of the mid-1990's, the impact of the Elite Guard still remains. Its positive traditions can be seen at numerous bases with Elite Gate Guard "sections". General Lemay most likely did not realize it, but he, successfully created one of the Greatest Security Forces traditions; all based on his vision.

Currently, at the former headquarters of Strategic Air Command, Security Forces still do much of the duties they were responsible for when General Lemay Commanded. They are still responsible for the Gates, locking the Commanders Office, Driving the Commander around, as well as providing personal protection for him. Although General Lemay wasn't a trained Security Policeman, he was/is the Greatest Air Force Cop.

Horse Patrol

The history of Horse use within the Air Force is as you can guess; minimal. The United States Air Force came of age in the late 1940's and early 1950's. The Army, Navy, and Marine Corps are much older services that didn't have the luxury of growing up in the era of the automobile. The Air Force, on the other hand, is based on vehicular (Aircraft) technology.

Horse Patrol within the Security Forces started at Clark Air Force Base in the Philippines in the mid 1980's. A horse unit was also initiated in Howard Air Force Base, Panama. Both Clark and Howard were prime candidates for Horse Patrols due to their overall environments of humid jungle. Horses can quietly and easily negotiate their way through terrain that few vehicles could manage.

Clark Air Force Base and its horse patrols were evacuated when the Volcano named Mount Pinatubo erupted after lying dormant for more than 500 years. Howard Air

Force Base would be returned to the Panamanian Government as per lease agreement along with the Panama Canal. Currently the base serves as a Multi-national Forces Base. The horses would trade the humid Panamanian climate for one just as good; California.

The Panamanian horses are currently patrolling at Vandenburg, Air Force Base in California. Vandenburg Security Forces are tasked with the protection of more that 30 miles of protected wildlife sanctuary. Horses are the logical choice of vehicle since they leave much less damage than any type of two or four-wheeled vehicle, including motorcycles or All Terrain Vehicles (quad ATV's).

Vandenburg horse patrols are also charged with environmental protection and enforcement of hunting regulations. Personnel assigned to this unit may be Federally Certified Game Wardens and usually carry 9mm handguns or shotguns in the commission of their duties.

Recently several more units are expressing interest in horse patrolling. The latest to succeed in creating a equine program is the 314th Security Forces Squadron located at Little Rock Air Force Base. They utilize the horse patrols in their effort to secure more isolated areas of their Arkansas base including hunting areas that were somewhat neglected prior to the horse patrols.

Alarms

Security Forces personnel are a true extension of a security system. Alarms need to be investigated, and this is only done by placing a pair of human eyes "on-target". Whether through cameras or responding troops, no alarm is terminated until security personnel arrive on-scene and investigate any possible causes of that alarm.

If the responding forces arrive on-scene and are engaged or incapacitated (killed), other responders have more information and will counter appropriately. However unfortunate, overall security of the base and its resources outweighs the value of any singular post or patrol and posts must be utilized accordingly.

With the prevalence of so many High Security and Restricted Areas placed within the confines of the majority of Air Force Base's and the limited resource of personnel, the need for automation arose. Alarm systems are a Force Multiplier and can be placed at strategic locations and allow constant and reliable monitoring of entrances to secured areas as well as aid in detecting unauthorized activity within or around facilities.

Installations such as the Pentagon could easily sap 1000's of personnel at numerous classified posts and doorways 24 hours a day. The use of alarms can give one individual the duty of monitoring an alarm panel for hundreds of areas, multiplying his singular effectiveness while boosting the effectiveness of overall security.

Alarms come in many forms and are used on every Air Base and Department Of Defense (DOD) facility the Security Forces Defend. From the low threat supply depot, to the most secretive and isolated facilities such as the Tonopah Test Range in the Nevada deserts.

Alarms monitor, intrusions of facilities along with openings and closings of doors, and hangers. Alarms can be registered by magnetic switches, infrared, radar, and even vibration sensors that can be buried under ground to detect approaching personnel and vehicles. All types of alarms are available to Security Forces to insure a high performance security capability exists, when utilized correctly.

But the fact remains and is worth mentioning once more, Security Forces are an extension of the security system and "no" post or patrol is more valuable than overall base security. The Air Force mission will always prevail.

Terrorism

1969 was an unofficial birth year for International Terrorism in all its forms, this is when highjacking became a beneficial tactic for those attempting to gain the worlds attention for their political causes. Terrorists realized highjacking was very cheap, easy, and made a huge political statement by focusing the world media's attention. Although many countries were isolated from such activities, this is when they would be introduced to the acts, as well as the people that would commit such acts.

Steps were required to combat this growing threat of Air Piracy. Security Police were doing their part on the ground by searching luggage at civilian airports with Military Working Dogs and training civilian police department dogs as well. They also found another means of aiding the civilian airline industry.

Grid Square

In September of 1970, President Richard Nixon would take the offensive on International terror and order that federal agents protect aircraft. 300 individual agents would come from the Federal Aviation Administration (FAA), US Marshal Service (USMS), Federal Bureau of Investigation (FBI), Department of the Treasury, as well as the Central Intelligence

Agency (CIA) would now serve and share duties as Sky Marshals.

Three Hundred Agents were better than the zero that were protecting the planes before the Sky Marshal program. Such a limited number would definitely not make much of an impact on the large volume of aircraft that were taking to the air on a daily basis. In October 1970, a combined force of 800 US Army Military Police and USAF Security Police were introduced to the project as an interim solution. This project was kept fairly quiet, and known unofficially by the name of GRID SQUARE.

Security Police assigned to the project were (somewhat legally) given the status of FAA agents despite the fact they were still under the jurisdiction of the Uniform Code of Military Jurisdiction (UCMJ). Normally military personnel cannot engage in civilian law enforcement duties because of Posse Comitatus[4]. GRID SQUARE authorized them to carry weapons on aircraft and make arrests on civilian personnel.

GRID SQUARE was the predecessor to todays Sky Marshal program and was in all accounts a successful mission. Although there were incidents where marshal's had to take actions to secure aircraft, there was never a need to use lethal force to defend the passengers or crew from terrorists attempting to seize control of any aircraft.

---

[4] Posse Comitatus Act of 1878 prohibits US military forces from arresting *(detaining is a separate legal status)* civilians, searching and seizing evidence or otherwise execute the laws while on U.S. Soil with the exceptions of few emergencies. **Title 18 U.S.C. 1385**

Since the mid 1970's, the Security Police have been cognizant of threats posed to USAF personnel and resources by terrorism. Even as the rest of the service tried to deny the threat and pay as little attention as possible, Security Police have maintained the focus and insured that they remain vigilant to their bases. They realize that although a big military base may not be a smart target for a terrorist, they may still become a target; and that only they were really watching.

Time and time again during the late 70's, terrorist would "shock the world" and America would "lose its innocence". Over and over again, acts of terrorism, due in part to Muslim Fundamentalism, would strike with devastating effect. By the early 1980's, Security Police had picked up a reputation as being a cunning and unpredictable security force. Many people believed it was due to their training, however, many others would argue that it was due to the flexibility afforded to the troop(s) in the area at the time, and a measure of luck.

Unlike many Army and Marine Corps guards of the day, Security Police were given an area and told to secure it. They failed to make patterns for any possible would-be observers to note and exploit. SP's were not where they were supposed to be, when they were supposed to be. If they were given a post and told to make hourly checks, they may or may not have been in their areas, or another post may be in their area with them. They may stop by in 2 hours, or they may stop by 10 minutes. These irregularities are not beneficial to an individual planning an elaborate attack. Attackers would simply to move to a softer target.

Recently, Security Forces have been initial responders to secure the sites of terrorist actions involving government, or military resources. Examples would include the bombings of the US Embassy's in the Kenyan and Tanzanian capitals as well as the USS Cole bombing in the port of Aden, in which a Navy ship was bombed leaving 17 Sailors dead and numerous others injured.

Khobar Towers

On June 25th, 1996, an Security Forces Staff Sergeant was checking on some of his posts placed atop of a dormitory building. He was in position to observe two men back up a tanker truck to the fence, hop out and run, only to jump into a waiting car and speed away. The Sergeant and his posts sounded the alarm by notifying his Central Security Control, and then supervised a couple of posts in evacuating the building

The Security Troops went floor by floor knocking on doors and attempting to warn people. Within four minutes of spotting the truck, the rolling bomb detonated, with a yield, by some calculations, of more than 25,000 pounds of TNT equivalent explosives. Nineteen US service members dead, and more than 500 others would be injured. The blast formed a crater that was approximately 16 feet deep and more than 85 feet wide in diameter.

The bomb would be described as the largest bomb used since the Gulf War and the biggest bomb used thus far by terrorists. This bomb was far larger than the device that exploded the Marine Barracks on 23 October 1983 in which 245 Marines, soldiers, and sailors were killed, with another 146 more wounded. Thankfully, on this day, distance from the detonation and some time to evacuate would limit the number of deaths.

Fortunately, the attackers realized they would never make it on the base without being detected upon entry. Obviously, due to the deterrent factor, the bombers must have felt their operation would be easier and more likely to succeed by avoiding Security Forces at the entry control point.

In the end, the blame has been passed around and no one is willing to accept any responsibility for anything that has happened, but, as far as Security Forces are concerned, Security was successful in completing its mission. The area the bomb was detonated in was previously noted as a major vulnerability. Not only had Security Forces noted this weakness prior and attempted to have more concrete barriers placed farther away only to have the request turned down by the Saudi government.

Security Forces are also credited with noticing the vehicle activity, sounding the alarm, and initiated all the actions that saved lives that dreadful day. This truth earned 4 personnel the Airman's Medal for their actions during this incident. One of those was a Surgeon and the remaining three individuals were Security Forces.

Raven

    Raven teams are Security Forces component personnel that provide security to aircraft by traveling along with aircraft while on their operational missions. An example of a Raven mission would be a humanitarian mission to a location to drop off food. The aircraft will be on the ground for 3 hours to unload, however there will be no US support at that location. A contingent of Security Forces personnel travel with the plane and provide close-in security while the aircraft is on the ground. Properly utilized Raven's represent a self-contained and prepared security force.

    Raven represents a slight change in the way security for aircraft is applied, however, this is not a new idea. Vietnam experience showed that Security Police traveled with aircraft and that it was very beneficial. As the aircraft landed, the aircrafts passengers could be screened prior to entry to the aircraft. This practice proved successful and effective in not only, deterrence of sabotage, but also in screening for unauthorized passengers, and detection of contraband.

Generally Raven missions placed security members in the uniform of the Aircrew, which includes flight suits and helmets. Flight suits are worn under all appropriate combat load-bearing vests, belts, and body armor. Weapons may vary from handguns, up to M-16 carbines, to shotguns or any combination of pistols and long guns.

Initially all Raven personnel were trained by the Air Force with an Air Mobility Command (AMC) course that involved everything an individual should know including tactics that would benefit aircraft security on an unimproved or contingency use airfield. However, signaling the death of an official mission, average Security Forces personnel can now be selected to "crew" on a Raven mission.

Contingency Cops

820th SFG

In March 1997, the 820th Security Forces Group was activated. It provides worldwide first-in force protection for Air Force contingencies. The 820th SFG also adopted the heraldry of the 82nd Combat Security Police Wing (CSPW), thus explaining the falcon's revival as the Security Forces emblem, with the addition of the motto "Defensor Fortis" on the scroll, which loosely translates to Defender of the Force.

The 820th group has all the capabilities to protect a base available on a short notice. Included are Heavy weapons capability, an Investigations section, Explosive Ordinance Disposal, as well as K-9 and training sections. Other resources that are brought to the fight by the 820th include vehicle support, and direct detachments of both the Air Force Office of Special Investigations (OSI) and Military Intelligence personnel.

As should be the case for the entire Air Force, 820th training takes a back seat to nothing. The emphasis to train is evident and goes right to the capabilities of the 820th. Many of the Security Forces make their way to Fort Benning for Airborne training (Jump School).

Other members are Air Assault qualified and ready to work from Helicopters. There are a few Rangers within the 820th that help insure expertise is maintained as well as to provide another layer of leadership.

Good equipment and resources have been allocated for use by the 820th in every area. Not simply in small items such as night vision goggles, weapons, or vehicles, but also in the priceless hand-selected personnel that strive to go through the application process.

The 820th Group is a great concept. There is a problem with this "project" however, and that is that no security unit of this nature has stood the test of time. SAFESIDE, Combat Security Police, SPEC's (Security Police Elements for Contingencies), and Ravens were/are all specialized security units used to showcase what increased training, preparation, and planning can do towards the AF's security mission. None of these programs are around today in the form that they were envisioned or designed.

It can be assumed if the 820th cannot find and maintain a legitimate uniqueness, or a specialized capability that few in the Air Force structure can do; they will fall into the same pattern as all regular Security Forces units.

The 820th must avoid the tendency to stop deploying as full squadrons (if it hasn't happened already), they will stop training as vigorously as planned, then stop deploying to backfill units that may be on site already, then stop deploying for contingencies, then stop being immediately deployable. The unit will then not be valuable and will be ended. That is security history and hopefully it will not happen. The 820th seems to be a step in the right direction.

The 786th SFS

The 86th Contingency Response Group was stood up as a rapidly deployable airfield operation unit. The Security Forces of the 786 SFS are capable of rapid deployment into crisis situations. They provide base security until follow-on forces can arrive and assume security duties from them.

The 786th has been utilized almost constantly since their inception in 1999 and is centrally stationed at Ramstein Air Force Base in Germany. They have deployed and set up security at numerous locations to include Kosovo, Albania, Romania, and Kyrgyzstan.

On 27 March 2003 members of the 786 took part in the largest Combat airborne assault since World War 2. The troops made the historic jump with more than 1200 members of the 173rd Airborne Brigade (Sky Soldiers) out of Italy. This combat jump landed the troops in Northern Iraq in support of Operation *IRAQI FREEDOM* and is an incredible testament to the versatility of Security Forces.

Counter Snipers

A move to provide Counter snipers has caught on and began to take root within AF Security. In 1993, the Air National Guard Security Forces initiated a course designed to counter the serious threat that snipers pose to air base defense operations. The course provides training in, not only marksmanship ,but also in the field skills that successful snipers need to carry out their duties.

In the course, not only is physical training stressed, but tactics, field craft, reconnaissance, and wind compensation are emphasized. The most important skill a counter sniper must hone is accurate range determination. Outstanding marksmanship cannot be achieved under combat conditions until all of these factors are in balance. The course tries to impart this into its students.

The largest factor built into the course is the stress placed on students. Exercises and missions must be successfully negotiated to complete the course and to stay in the program. The counter sniper is always augmented with a spotter. Spotters aid in range determination, target identification as well as every other task required until the moment of trigger pull. All students must be proficient in both shooter and spotter positions.

Rifles included in the training include not only the M-16, but also the M-24 and the M-82 rifles. The M-24 is a bolt-action rifle that fires the 7.62 mm round. This is the same *caliber and type* of round that is fired from the M-60 machine gun. The M-24 carries a 5-round internal magazine. The action is based on that of the Remington 700, which has been around and improved for more than 4 decades.

The M-24 rifle actually represents the return of the bolt-action rifle to the Army Sniper program from the Semi-automatic variants of the M-14. One could imagine that it must be a very effective platform. Students must successfully engage targets from as close as 300 meters and out to 1000 meters.

Students are also familiarized with the M-82A1 .50 Caliber rifle system. The "light 50" fires the same size round as the M-2, .50 machine gun. This is usually reserved to engage hard targets, or the targets that the average rifle will have no effect on, including Radar systems, bunkers, and even medium armor targets. Maximum effective range for the M-82 is 1843 meters; however, larger targets out past 2000 meters can be engaged with devastating effect.

The counter sniper role proves to be valuable since it counters one of the most deadly types of foot soldier. Throughout history, the single sniper has wreaked havoc on everyone from lone individual targets, to platoon size units and even hampered operations on entire military bases by their presence alone.

The course has been so beneficial, that other US services have sent personnel for training. Most importantly, the Army has granted the course full and formal accreditation. The counter sniper course has proven to be, and will continue to be, successful.

A Counter-Sniper first

For the first time since the Counter-Sniper program began, two Security Forces went on an "offensive" mission with members of the Army's legendary 82nd Airborne. The mission included a long-range reconnaissance patrol. The patrol marks the first time Security Forces have been utilized in this fashion with the Army. Contact was made with enemy units on two occasions. The patrol came under distant attack in the form of rocket propelled grenades (RPG) or mortar fire, no casualties were recorded and unfortunately, no shots were fired by the counter-snipers to make for a more dramatic story.

The Marksmen were members of the 820th Security Forces Group out of Moody Air Force base, Georgia. The patrol itself was in support of *OPERATION ENDURING FREEDOM* in Afghanistan.

Special Ops Cops

There are an elite few Security Forces who are chosen based on their performance and motivation that operate within the Special Operations community. Their missions are diverse, many are CLASSIFIED, and they work directly in support of Special Operations Forces. Organization of units is not based on the typical USAF Squadron, Flight, Squad elements, instead they work In Detachments. Joint Special Operations Command (JSOC) deploys these cops out of Macdill and Pope Air Force Bases.

Training for these "Cells" must include a stint at Army Airborne School to learn to parachute. Army Special Reaction Training (SRT) course is the Army's SWAT School that must also be attended. Many operations require the protection of VIP's so the DOD

Protective Services Course is another course that must be completed.

Other courses that must be completed include the Mountain and Winter/Arctic Survival. Many other Special Ops Cops will continue on to complete Ranger School during their assignments. Just as Special Operators do, these few are amongst the most frequently deployed personnel within the Security Forces.

Along with these Detachments, there are other Security personnel and units that are regularly tapped for use by the Joint Special Operations Command (JSOC) and this includes the Security Forces of Hurlburt Field, Florida. Hurlburt Field is the home of Air Force Special Operations Command (AFSOC). The Security Forces unit there has also picked up the nickname of the "Special Ops Cops".

A Culture of Cops

Security personnel within the United States Air Force have always been different than other Airmen. In most cases, security troops cannot relate to the general AF so they isolate themselves. Since the early 1950's, the perception within the security community is that Army has more in common to the Security Forces than other USAF specialties.

The Army and Marine Corps "build" Soldiers or Marines. AF Basic Training "prepares" people to be specialists and not Airmen. Marines will say, "I'm a Marine", while Airman will say, "I'm a firefighter" or a

"Loadmaster". Extreme pride and esprit de corps that Soldiers or Marines attain after their completion of Basic Training is foreign idea most Air Force personnel.

Security personnel used to exhibit more pride and esprit de corps after Security Forces Training, however, this has been waning in past few years. Likely causes include, (1) the numbers game to place people in Security Forces despite their lack of motivation, ability and interest and, (2) the attempt to ensure *everyone passes* when obviously, *everyone should not pass*

Air Force personnel tend to stay away from security troops based on stereotypes that they are aggressive and that they only associate with other Cops. The truth is, although Security Forces have a job that may not be as technical as other Air Force specialties, and they may not have it as good, generally they think they are better than everyone else.

The average Security Troop *believes* that he generally puts in longer hours under more austere, and sometimes abrasive working conditions (by USAF standards), with unit leadership that is not always listening. Security Forces believe they have traits that other organizations within the Air Force lack.

Not only have Cops segregated themselves mentally, they have also stood out while in uniform. Within the Air Force, Cops were the first trained to wear their uniform trousers bloused (with pants tucked up into boots) as the Marines and Army did. They were the first to rid themselves of all green fatigues in favor of the camouflaged Battle Dress

Uniforms (BDU) and they will likely be amongst the first to move into a better Combat BDU.

Security Forces still have a pride within the Air Force of having the sharpest looking uniform preparation. While many units suggest uniforms must fit the minimum standard of "clean, dry, and serviceable"; those same Air Force personnel assume cops will be starched, pressed, and have a high shine on their boots.

Military Police First

Security Forces are unique in the organization of the Air Force. The Military Police role is vital to the security and law enforcement of every military installation. Security is a mission of the military that has been required for thousands of years. Whether in the Armies of Napoleon, or even in the form of the Roman Legion's Praetorian Guards, security is and will continue to be an essential part of the United States Air Force as well.

The future of Security Forces must include long-term goals, vision, and azimuth as directed from the USAF Top Cop just as much as the Marines have the Commandant of the Marine Corps. It is referred to as Leadership!

Security Forces must refocus on their efforts on their ultimate duty, which is to take charge of the post until properly relieved (General Order Number One). Not only should Security Forces be effective at base level duties, they must be interoperable and deployable with

Military Police of the sister services. It must also strive to be the best Department of Defense Security Force while never forgetting that they were, are, and will continue to be Military Police First.

Over and Out!

Security Forces Timeline

1941 June	United States Army Air Corps established.
1941 February	53,000 blacks allowed to enter the Army Air Forces.
1941 November	The US Army Air Force established black Air Base Security Battalions (ABS), for air base defense
1943	All black ABS Battalions successfully see combat in the North African and Mediterranean Theaters
1947 September	The US Air Force becomes a separate American military service from the US Army
1948	Air Force Military Police became Air Police (AP) in the newly created Air Force
1949 July	The U.S. Air Force becomes the first service to end racial segregation in its ranks
1950	Kimp'o Air Base (SAC), So. Korea, strafed by North Korean fighters. As a result, Colonel James Luper, Air Provost Marshal, is ordered by General Curtis LeMay to build a ground combat force for the Strategic Air Command (SAC). This force was the predecessor of Air Base Defense and was nicknamed "Lupers Troopers".
1950 June	The Korean War starts and Air Police expand in size from 10,000 troops to 39,000 by December, 1951.
1950 September	Air Police School training was established at Tyndall AFB, Florida.
1951	Strategic Air Command (SAC) Security School for Air Base Defense is placed at Fort Carson, CO.
1953	USAF Air Base Defense school established at Parks AFB, California, for regular AP units.
1956 October	Air Police School moved to Lackland AFB, Texas
1956	USAF opens Military Working Dog (MWD) training centers in Germany and Japan.
1957	Air Force assumes training of all sentry dogs for the Department of Defense at Lackland AFB, Texas.
1959	First "official" AF Air Police Shield authorized although there were unofficial versions prior.
1960 January	Small arms training is turned over to the Base Marksmanship School (Red Caps).
1961	Missile security was formally established with the first Titan Missile sites at Davis Monthan AFB, AZ., McConnell AFB, KN., Little Rock AFB, AR., and Malstrom AFB, MT.
1961	"Operation Farmgate." Air Policemen deploy to guard US cantonment areas at Tan Son Nhut AB, Bien Hoa AB, and DaNang AB, Vietnam.
1962	.38 caliber revolvers phased in to replace the aging .45 caliber pistol fleets (SAC units received them first).
1963 December	USAF had one Air Police officer and 280 enlisted on TDY status in Vietnam. One out of every 16 new recruits became an Air Policeman.

171

1964 June	M16 rifle with 5.56mm ammo phrased in to replace M1 and M2 carbines.
1965	Change from Air Police into Security Police to reflect new emphasis on security and policing.
1965	Operation Safeside initiated; the 1041st begins personnel selection and training
1965	Vietnam Manpower requirements for southeast Asia jumped from 148 to 2,880.
1965	SSgt Terrence Jensen is first AP killed in Vietnam. SSgt Jensen was killed on July 1 while flushing out Vietcong sappers at DaNang AB. This incident brought MWD teams to Vietnam , and by 1966, not one Air Base had been penetrated by bases guarded with MWD's.
1965 September	Air Police squadrons receive their first armored vehicles in the form of the Cadillac Gage Commando.
1966	Air Police strength for southeast Asia increased to more than 5,000 Security troops on the ground.
1966	Special Order G42, July 18,1966, activated the 1041st Security Strike Force Squadron (Test) commanded by Lt Col William Wise and gave birth to Operation Safe Side. This unit was Ranger trained to provide an aggressive role in air base defense. Later they were renamed the 1041st Security Police Squadron (Test).
1966	Light blue beret with falcon emblem authorized for use for combat Security Police assigned to 1041st SPS(T), visible identifier of their role as a specialized unit.
1967	1041st SPS(T) arrived at Phu Cat AB, Vietnam.
1968	Tet Offensive: Division strength NVA and Viet Cong hit the fences at Ton Son Nhut, Bien Hoa, DaNang, and Tuy Hoa Air Bases. Attacks continued through February, but the security police held their ground; not one air base was lost to the enemy.
1969 August	First official patrol dog training class began. Sentry (attack) dog training was phased out.
1971	Security Police career field split in two, Security Specialist and Law Enforcement Specialist
1971	Women allowed to enter the Law Enforcement career field and were authorized to wear a white beret.
1971 November	First six women graduated from the Law Enforcement Course (Sgt. Biggs, A1c Foster, A1c Byers, A1c Heims, and A1c Hollingsworth).
1973	First female SP commissioned officer, Lt Sally Kucera, graduates Officer Academy School.
1973	First female graduated from Security Police Officers Course, Lt Sally Kucera (Col. Sally Uebelacker).
1973 December	First females entered Patrol Dog Handler's Course and graduated in 1974. (Airmen Shiela Dugan and Rickie Thompson)
1975	Authorization of SP Qualification Badges, dark blue beret, white scarf, and blue short jacket.
1975	320 SP's direct the evacuation of Saigon during the airlift. Air Force police units were the first in and the last to leave Vietnam.

1975	Major General Sandler, Tom becomes the first Security Police General Officer.
1976 March	First three female Security Police Officers graduates from ABGD III Course (1 Lt Pamila Krauss, 2 Lt Noreen Aberico, and 2 Lt Patricia Schafer).
1976 October	18 month test program begins for women in Security Specalist career field. 100 women enter program and are sent to 4 bases (Barsdale AFB, Grand Folks AFB, Nellis AFB, and Osan AB, Korea). At the end of the test, career field remains closed to women due to sexism in training. Participants in the program were allowed to retrain or separate from the Air Force.
1977	Security Police Museum opens at Lackland Air Force Base, Texas.
1979	Emergency Services Team (EST) school established, Lackland AFB, Texas.
1982	Combat Arms Training, Maintenance is put under he operational control of Security Police units.
1987	Fort Dix, NJ becomes the home of Ground Combat Skills as the Army takes over the training of Air Base Ground Defense (ABGD)
1990 August	Operation Desert Shield sends in the first Security Police units to Saudi Arabia (prior only in Egypt, Saudi Arabia)
1995 August	Ground Combat Skills becomes an Air Force course once more at Camp Bullis in Texas.
1996 June	Security Forces detect the Khobar Towers terrorists and initiate evacuation just before terrorists detonate a truck bomb in Saudi Arabia
1996 July	Security Forces are immediately on-scene to the Olympic bombing of Centennial park while acting in the largest multi-organization, civil security force in US history
1997 October	Career field name officially changes from Security Police to Security Forces. Security Specialists, CATM, and Law Enforcement Specialists incorporate
1997	Cloth "Defensor Fortis" flash replaces MAJCOM beret crest.
1998 August	Security Forces arrive on-scene within 12 hours of terror bombing of U.S. Embassy's bombing in Kenya, Africa
2000 October	Security Forces are amongst first troops to respond to the USS Cole after it was attacked by terrorists in the port of Aden, Yemen.
2003 March	786th Security Forces jump into Iraq during Operation Iraqi Freedom. First Security Forces combat jumps since the Vietnam war.

Security Forces must-haves:

Lemish, Michael G. *War Dogs, A History of Loyalty and Heroism,* Virginia: Brassey's, 1999.
An in depth look at the entire history of the Military Working Dogs from their origins. The Department of Defense program is examined to

Fox, Lieutenant Colonel, Roger P. *Air Base Defense in the Republic of Vietnam 1961 1973.* Washington, DC.: Office of the Air Force History, 1979.
An entire look of the history of the Security Police. Years included are from 1961 through to 1973.

The Diagram Group, *WEAPONS; An international encyclopedia from 5000 BC to 2000* New York: St. Martin's Press 1990.
Leader in photos, descriptions, and other information on weapons from the beginning of recorded time to the present.

SAFESIDE, United States Air Force, Film Report 822, 1968. Air Force Videotape.
Video History of the initial Safeside units including training, resources, equipment, alarms, weapons. Also gives history of the reasons that that the program began, as well as stats of Aircraft lost, and base attacks

Vick, Alan; *SNAKES IN THE EAGLES NEST, Ground Attacks on Air Bases, 1940 992.* RAND Corporation
A researched look at Ground Attack Data and the effectiveness of air base ground defense of air bases. Great read with plenty of insight into why attacks succeeded or failed.

Photos

Thanks to all those who supported me by sending pictures or pointing me in the right direction to get them. What look into history would be complete without lots of captured moments. Many photographs were gathered, collected or donated from the public domain. PD photos were taken by individuals for use by their various government agencies, magazines, websites, historian offices, or Military Organizations. Other photos were provided by individuals who immortalized moments during the courses of their military careers. A few pictures have unknown origins. With no way of tracking that origin, the decision was made to utilize the pictures due to their historic value.

Donate Information

More information or history, or pictures can be sent to the author at krpinckney@hotmail.com. Thanks you much and together we can provide a more informational Second Edition.

DISCLAIMER

THIS BOOK IS NOT AN OFFICIAL UNITED STATES AIR FORCE WORK. IT IS A BRIEF HISTORICAL LOOK INTO THE USAF SECURITY FORCES. THERE ARE OPINIONS EXPRESSED HEREIN, AND THEY ARE THE AUTHORS' OWN. THEY DO NOT NECESSARILY REFLECT THE VIEWS, OPINION, OR THE POSITION OF THE UNITED STATES AIR FORCE, THE US DEPARTMENT OF DEFENSE OR THE UNITED STATES GOVERNMENT IN ANY WAY, SHAPE, OR FORM.

About the Author

Kali Pinckney enlisted into the United States Air Force after graduating high school in 1992. He was trained and served as a Security Specialist in various units, in various locations within the United States and overseas to include panama and Cuba. His duty's included Aircraft Security, Prisoner Transport, Air Base Ground Defense, Armory, Hostage Negotiations, as well as Nuclear Security.

Making the decision to separate in 1996, he enlisted in the 144th Security Forces Squadron, of the California Air National Guard located in Fresno California. Upon gaining a Commission from the Air Force Reserve Officer Training Corps in 2000, he re-entered the USAF once again, this time as a Security Forces Officer. He holds a degree in Criminology from California State University, Fresno.

Printed in the United States
42295LVS00002B/130-132